This book presents the integrated contributions of hydrologists, meteorologists and ecologists to the first International Hydrology Program/International Association of Hydrological Sciences (IHP/IAHS) George Kovacs Colloquium in connection with the study of global hydrology and climate change. All the atmospheric, hydrological, and terrestrial components of the Earth's systems operate on different time and space scales. To resolve these scaling incongruities requires a model of the complex interactions of land surface processes at the different scales. This represents a major challenge for hydrologists, ecologists and meteorological scientists alike. This book deals with time and space scale variations with reference to several topics; including soil water balance, ecosystems and interaction of flow systems, and macroscale hydrological modelling. This book will be of great use to researchers, engineers and forecasters with an interest in space and time scale variability.

T0238936

Space and Time Scale Variability and Interdependencies in Hydrological Processes

INTERNATIONAL HYDROLOGY SERIES

Space and Time Scale Variability and Interdependencies in Hydrological Processes

Edited by
Professor Reinder A. Feddes

(Department of Water Resources, Wageningen Agricultural University, The Netherlands)

CAMBRIDGE UNIVERSITY PRESS
Cambridge, New York, Melbourne, Madrid, Cape Town, Singapore, São Paulo

Cambridge University Press
The Edinburgh Building, Cambridge CB2 2RU, UK

Published in the United States of America by Cambridge University Press, New York

www.cambridge.org
Information on this title: www.cambridge.org/9780521495080

First published 1995
This digitally printed first paperback version 2005

A catalogue record for this publication is available from the British Library

Library of Congress Cataloguing in Publication data
Space and time scale variability and interdependencies in hydrological
processes / edited by Reinder A. Feddes.
     p. cm.
    Includes bibliographical references and index.
    ISBN (invalid) 0-521-49508-3
    1. Hydrology – Congresses.   2. Climatic changes – Congresses.
  3. Space and time – Congresses.   I. Feddes, R. A.
  GB652.S67   1995
  551.48 – dc20   94-42930   CIP

ISBN-13 978-0-521-49508-0 hardback
ISBN-10 0-521-49508-3 hardback

ISBN-13 978-0-521-02293-4 paperback
ISBN-10 0-521-02293-2 paperback

Contents

REMOTE SENSING – INVERSE MODELLING APPROACH TO DETERMINE LARGE SCALE EFFECTIVE SOIL HYDRAULIC PROPERTIES IN SOIL–VEGETATION–ATMOSPHERE SYSTEMS
R.A. Feddes

THE IMPORTANCE OF LANDSCAPE POSITION IN SCALING SVAT MODELS TO CATCHMENT SCALE HYDROECOLOGICAL PREDICTION
T.J. Hatton, W.R. Dawes and R.A. Vertessy

THE INFLUENCE OF SUBGRID-SCALE SPATIAL VARIABILITY ON PRECIPITATION AND SOIL MOISTURE IN AN ATMOSPHERIC GCM

D. Entekhabi

MODELLING THE HYDROLOGICAL RESPONSE TO LARGE SCALE LAND USE CHANGE

A. Henderson-Sellers, K. McGuffie and T.B. Durbidge

AN APPROACH TO REPRESENT MESOSCALE (SUBGRID-SCALE) FLUXES IN GCMs DEMONSTRATED WITH SIMULATIONS OF LOCAL DEFORESTATION IN AMAZONIA

R. Avissar and F. Chen

A HIERARCHICAL APPROACH TO THE CONNECTION OF GLOBAL HYDROLOGICAL AND ATMOSPHERIC MODELS
G.W. Kite, E.D. Soulis and N. Kouwen

STOCHASTIC DOWNSCALING OF GCM-OUTPUT RESULTS USING ATMOSPHERIC CIRCULATION PATTERNS
A. Bárdossy

DEPENDENCIES OF SPATIAL VARIABILITY IN FLUVIAL ECOSYSTEMS ON THE TEMPORAL HYDROLOGICAL VARIABILITY
H.P. Nachtnebel

PROBLEMS AND PROGRESS IN MACROSCALE HYDROLOGICAL MODELLING
A. Becker

PREDICTABILITY OF THE ATMOSPHERE AND CLIMATE: TOWARDS A DYNAMICAL VIEW
C. Nicolis

FROM SCALAR CASCADES TO LIE CASCADES: JOINT MULTIFRACTAL ANALYSIS OF RAIN AND CLOUD PROCESSES
D. Schertzer and S. Lovejoy

FRACTALS ET MULTIFRACTALS APPLIQUÉS À L'ÉTUDE DE LA VARIABILITÉ TEMPORELLE DES PRÉCIPITATIONS
P. Hubert

Acknowledgement

The Scientific Editor expresses his sincere appreciation to Ir. H. Salz, former secretary-general of the International Institute for Infrastructural, Hydraulic and Environmental Engineering, Delft for his careful technical editing and Mrs. G.J. Quint for her precise and fast typing of the manuscripts.

Opening address of the first George Kovacs colloquium

U. SHAMIR, President IAHS

Faculty of Civil Engineering,

Technion-Israel Institute of Technology

3200 Haifa,

Israel

1. GEORGE KOVACS

It is a privilege and a pleasure to open this George Kovacs Colloquium convened jointly by UNESCO/IHP and IAHS.

UNESCO and IAHS have named this series of Colloquia after George (Gyorgy) Kovacs, a renowned international hydrologist, and President of IAHS from 1983 to 1987, who passed away in 1988. This first IHP/IAHS George Kovacs Colloquium is devoted to Space and Time Scale Variability and Interdependencies in Various Hydrological Processes.

George Kovacs was a talented and energetic innovator in the fields of groundwater, hydrology and hydrogeology, and a researcher of broad perspective. His vision and leadership brought him to the pinnacle of the scientific community in Hungary and beyond. He had a leading role in the international hydrologic community, and it is most appropriate that we dedicate this Colloquium to his memory.

2. COOPERATION BETWEEN UNESCO AND IAHS

UNESCO and IAHS have enjoyed a long standing co-operation. IAHS serves in the capacity of science advisor to UNESCO's Division of Water Sciences, provides input to UNESCO and its International Hydrological Programme (IHP) through ideas and proposal of experts to serve on UNESCO water project committees and task forces. IAHS has helped UNESCO in setting up the group of experts who are presently considering IHP-V, the next phase of IHP, to be titled: 'Hydrology and water development in a vulnerable environment'.

3. SPACE AND TIME SCALES AND VARIABILITY

Space and time scales and variability are fundamental issues in scientific and operational hydrology. Variability in time and space, and the question of whether we can transport results from one scale to the other, are basic to all our work.

The early development of hydrology did not address this issue explicitly, but it was always there, at the core of research and application. Consider the early tools of modern hydrology, for example the unit hydrograph. The *watershed* was lumped, and assumed to be a linear system. The processes of precipitation, infiltration, evaporation, runoff, and channel routing were all folded into a single aggregate representation of the watershed.

As a result of this aggregation and the assumption of linearity it became possible to compute useful results, and to investigate the difference between watersheds in various regions and conditions. The validity of the approach was predicated on the fact that computed unit hydrographs fit reasonably well the event from which they were derived.

After such techniques had been used quite effectively for many years, research began to focus on the validity of the underlying assumptions, in particular those that relate to spatial and temporal distribution, aggregation and averaging of all the phenomena in *water systems*. Distributed models were developed and employed, as our ability to solve them, analytically and especially numerically, improved, and the relation of phenomena described at various scales gained importance.

In parallel, *groundwater hydrology* faced issues of a somewhat similar nature. The partial differential equations for flow and transport in porous media are based on averaging

1

over a 'representative elementary volume' (REV), which is small enough to be considered as a point in the field, yet large enough to be an average over grains and pore space. In numerical solutions of field problems there is a need in representative values for much larger cell sizes. A basic questions arises: do we use the same partial differential equations, for example the convection-dispersion equation, at all scales, from a few centimetres to thousands of metres, adjusting the coefficient values so the model replicates field data, or are there intrinsically different phenomena as we move from one scale to the next? And another question of practical importance in groundwater hydrology is: when is it admissible to use an average over the entire depth of an aquifer as a representative value for a point in a two dimensional model?

The stochastic approach to groundwater flow and transport gets around some of these questions in a different way, by allowing the probability functions to reflect the effect of scale. But even here, some assumptions about spatial behaviour underlie the approach. The question remains of whether the phenomena are truly random, or is the stochastic approach merely a way of substituting our ignorance of the underlying physics and/or a way of getting around the difficulty in obtaining the true values of physical properties?

Scientific and operational hydrologists are interested in problems which range from point processes, through plots and small fields, watersheds over a wide range of sizes, to regional, continental, and global scales. It is therefore useful to determine whether processes at these various scales are self-similar, because if they are, then we should be able to transport results from one scale to another. Heterogeneity increases with scale. Is it only because over larger distances and times, there is more chance for larger variations, or are the underlying laws different? Answering this question determines the meaning of averages and other representative values. It also provides guidelines for determining the spatial and temporal intervals for data collection.

More recently interest in time and space scale and variability has resulted from the need to have a *better integration of meteorology* and *hydrology*, in connection with the study of global hydrology and climate change. GCMs operate at a scale which is considerably larger than normally used in watershed hydrology, and therefore interfacing between GCMs and terrestrial models is difficult.

More generally, combined work on different components of the earth system requires bridging across scales, in space and/or time. The interface between these components of the global system is in mass, momentum, and energy transfers. It is therefore necessary to aggregate and disaggregate these quantities across the common boundaries, so that the subsystems can be studied jointly.

Ecological modelling usually requires a much finer spatial scale. Here the perspective is local to regional, although some global issues also emerge.

Space and time scale variability and interdependencies are equally important for *physical, chemical, and biological processes*, except that they are probably even more complex than for water quantity. Variability ranges from the molecular and living cell scales, through 'control volume' sizes, to a river cross section and a river reach, lumped and distributed models of lakes and reservoirs, horizontal and vertical distribution of constituents in aquifers, and all the way up to the regional scale. The times of interest also range between seconds and minutes, up to multi-year periods and sometimes many decades. Research is needed to provide us with an understanding of scale and variability in these processes, with the foundation for determining what are good representative values, and how monitoring of water quality should be carried out.

4. NEW TECHNIQUES

Finally, a word about some new approaches and techniques. Fractals are a means for studying phenomena which are selfsimilar at different scales. Fractals should therefore be a useful device in the topics addressed by this Colloquium. The theory of chaos is suggested as more than just a clever tool. It has been said that chaos is the next most important scientific discovery after relativity and quantum physics. This may be so, but even if that is an exaggeration, chaos theory and tools should provide some useful ways of dealing with variability. As is sometimes the case, fractals and chaos may create a fad which exceeds the proper perspective for these techinques. One is therefore well advised to study and apply these techniques, but not present them as more than they actually are.

ACKNOWLEDGEMENTS

In conclusion, I would like to thank the Division of Water Sciences of UNESCO, its Director Dr. Andras Szollosy-Nagy and his staff, the invited speakers at this Colloquium, and all who assembled here for this event, in which we honour the memory of George Kovacs, a Past President of IAHS, and study together one of the more important and exciting aspects of hydrology.

Heterogeneity and scaling land-atmospheric water and energy fluxes in climate systems

E.F. WOOD

Department of Civil Engineering and Operations Research

School of Engineering and Applied Science

Princeton University, Princeton

New Jersey 08544, USA

ABSTRACT The effects of small-scale heterogeneity in land surface characteristics on the large-scale fluxes of water and energy in the land-atmosphere system have become a central focus of many of the climatology research experiments. The acquisition of high resolution land surface data through remote sensing and intensive land-climatology field experiments (like HAPEX and FIFE) has provided data to investigate the interactions between microscale land-atmosphere interactions and macroscale models. One essential research question is how to account for the small-scale heterogeneities and whether 'effective' parameters can be used in the macroscale models. To address this question of scaling, three modeling experiments were performed and are reviewed in this paper. The first is concerned with the aggregation of parameters and inputs for a terrestrial water and energy balance model. The second experiment analyzed the scaling behaviour of hydrological responses during rain events and between rain events. The third experiment compared the hydrological responses from distributed models with a lumped model that uses spatially constant inputs and parameters. The results show that the patterns of small scale variations can be represented statistically if the scale is larger than a representative elementary area scale, which appears to be about 2–3 times the correlation length of the process. For natural catchments this appears to be about 1–2 km^2. The results concerning distributed versus lumped representations are more complicated. For conditions when the processes are non-linear, lumping results in biases; otherwise a one-dimensional model based on 'equivalent' parameters provides quite good results. Further research is needed to understand these conditions fully.

1. INTRODUCTION

The complex heterogeneity of the land surface through soils, vegetation and topography, all of which have different length scales, and their interaction with meteorological inputs that vary with space and time, result in energy and water fluxes whose scaling properties are unknown. Research into land-atmospheric interactions suggests a strong coupling between land surface hydrological processes and climate (Charney et al., 1977; Walker and Rowntree, 1977; Shukla and Mintz, 1982; Sud et al.,1990.) Due to this coupling, the issue of 'scale interaction' for land surface-atmospheric processes has emerged as one of the critical unresolved problems for the parameterization of climate models.

Understanding the interaction between scales has increased importance when the apparent effects of surface heterogeneities on the transfer and water and energy fluxes are observed through remote sensing and intensive field campaigns like HAPEX and FIFE (Sellers et al., 1988). The ability to parameterize macroscale models based on field experiments or remotely sensed data has emerged as an important research question for programmes such as the Global Energy and Water Experiment (GEWEX) or the Earth Observing System (EOS). It is also important for the parameterization of the macroscale land-surface hydrology necessary in climate models, and crucial in our understanding how to represent subgrid variability in such macroscale models.

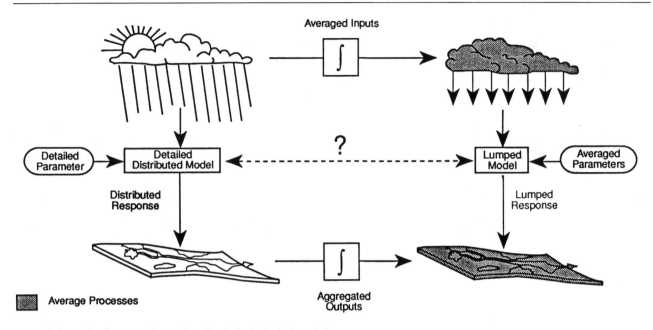

Fig. 1 Schematic of aggregation and scaling in hydrological modeling.

From a modeling perspective, it's important to establish the relationship between spatial variability in the inputs and model parameters, the scale being modeled and the proper representation of the hydrological processes at that scale. Fig. 1 presents a schematic for modeling over a range of scales.

Let us consider Fig. 1 in light of the terrestrial water balance, which for a control volume may be written as:

$$\left\langle \frac{\partial S}{\partial t} \right\rangle = \langle P \rangle - \langle E \rangle - \langle Q \rangle \qquad (1)$$

where S represents the moisture in the soil column, t is time, P the precipitation from the atmosphere to the land surface, E evaporation from the land surface into the atmosphere and Q the net runoff from the control volume. The spatial average for the control volume is noted by $\langle \bullet \rangle$.

Eq. (1) is valid over all scales and only through the parameterization of individual terms does the water balance equation become a 'distributed' or 'lumped' model. By '*distributed*' model, we mean a model which accounts for spatial variability in inputs, processes or parameters. This accounting can be either *deterministic*, in which the actual pattern of variability is represented – examples include the European Hydrological System model (SHE) (Abbott et al., 1986a, b) and the three-dimensional finite element models of Binley et al. (1989) or Paniconi and Wood (1993); or *statistical*, in which the patterns of variability are represented statistically – examples being models like TOPMODEL (Beven and Kirkby, 1979) and its variants (see Wood et al, 1990; Famiglietti et. al., 1992; Moore et al, 1988; Wood et al. 1993) in which topography and soil plays an important role in the distribution of water within the catchment.

By a '*lumped*' model we mean a model that represents the catchment (or control volume) as being spatially homogeneous with regard to inputs and parameters. There is a wide number of hydrological water balance models of varying complexity that don't consider spatial variability. These range from the well-known unit hydrograph and its variants, the water balance models of Eagleson (1978), to complex atmospheric-biospheric models being proposed for GCMs (examples being, the Biosphere Atmosphere Transfer Scheme (BATS) of Dickinson (1984) and the Simple Biosphere Model (SiB) of Sellers et al. (1986).

The terrestrial water balance, including infiltration, evaporation and runoff, has been revealed to be a highly nonlinear and spatially variable process. Yet, little progress has been made in relating the observed small-scale complexity that is apparent from recent field and remote sensing experiments to models and predictions at large scales. It is this relationship that is the subject of this paper. The research being presented incorporates recent work in investigating the effects of spatial variability and scale on the quantification and parameterization of the terrestrial water balance. The results draw primarily from the papers of Wood et al. (1988, 1990), Wood and Lakshmi (1992), and Famiglietti and Wood (1995). Important related papers are those of Wood et al. (1986), Sivapalan et al. (1987), Beven et al. (1988) and Beven (1988).

2. CHANGING SCALE AND WATER BALANCE FLUXES

Large-scale field experiments such as FIFE and HAPEX, and remote sensing experiments like MAC-HYDRO (see Wood et al., 1993) and MAC-EUROPE (see Lin et al., 1993), have shown the significant variability across a catchment with regards to runoff production, soil moisture levels and actual evaporation rates. The heterogeneity in hillslope forms, soil properties and vegetation combine with variability in rainfall to produce different runoff processes and responses across hillslopes, different soil moisture conditions and interstorm (dry period) moisture redistribution and evapotranspiration.

For a hillslope, it may be possible to develop a distributed model which explicitly considers variability in soil and vegetation properties. In fact, the simulations of Smith and Hebbert (1979) show that the actual patterns of soil properties may be important in simulating the runoff response from a hillslope.

At the scale of a small catchment, it may be possible to consider the variabilities in topography, soil and vegetation as if they came from a stationary statistical distribution (Beven, 1988; Wood et al., 1990.) Thus the distributed model would consider patterns of variability statistically. Within a physioclimatic region, we can consider that there may be a population of small catchments that are statistically similar but whose actual patterns of topography, soil and vegetation properties and therefore responses vary quite differently (Beven, 1988).

As scale increases, so does the sample of the small catchments and therefore the sample of the properties that control the water balance fluxes. This increased sampling of small catchments leads to a decrease in the difference between small catchment responses, even though the patterns of the properties are quite different across these small catchments (Beven, 1988; Wood et al., 1990). At some scale, the variance between the hydrological responses for catchments (or areas) should reach a minimum. Wood et al. (1988) suggested that this threshold scale be referred to as the 'Elementary Representative Area' (REA) which they define as: *the critical scale at which implicit continuum assumptions can be used without explicit knowledge of the actual patterns of topographic, soil, or rainfall fields. It is sufficient to represent these fields by their statistical characterization.*

Predicting the water balance at the REA scale may very well require considering heterogeneity at smaller scales, through its statistical characterization; it should not imply the use of equivalent and average parameters. In terms of Fig. 1, changing scale helps us understand the aggregation of the output from the distributed response. The concept of the REA scale helps us in clarifying the relationship between a distributed model and the lumped model, and how this relationship may vary with scale.

In this paper we report on a series of numerical experiments that investigate aggregation and scaling of land-surface hydrological processes. Famiglietti (1992), Famiglietti et al. (1992) and Famiglietti and Wood (1994a) have developed a water and energy balance model within a TOPMODEL-like structure that predicts water and energy balance fluxes for areas of heterogeneous soil, hillslopes, rainfall and net radiation characteristics. The models are summarized in Appendix A, and were developed to predict water and energy fluxes for the Intensive Field Campaigns (IFCs) of FIFE (Famiglietti and Wood, 1994a, b) and subsequent remote sensing experiments (Wood et al., 1993; Lin et al., 1994). The models have also been used to analyze the water balance fluxes for catchments of different scales, in which the small catchments were sampled from a particular topography – in this case the topography of the FIFE area (Famiglietti and Wood, 1995).

The *numerical experiments* that will be reported here are as follows. The first is the aggregation of distributed inputs for the water balance model; specifically the representation of soil and topography, and vegetation. The second is the aggregation of the hydrological responses in a catchment due to rainfall during a storm event and due to evaporative demands during interstorm periods. These two sets of experiments allow us to infer the nature of aggregation in parameters and processes. The third experiment will compare the aggregated fluxes from the distributed model to the predicted fluxes from a lumped version of the model.

3. CHANGING SCALE AND MODEL INPUTS

3.1 Scaling of topography

Appendix A provides a summary of the water and energy balance models. The models were applied to the Kings Creek catchment in the FIFE area in Kansas. The FIFE area is 15 km x 15 km, with a rolling topography with an approximate elevation range of 325 m to 460 m. Except for heavier vegetation at the bottom of stream valleys, the vegetation consists of short crops, pasture and natural grasses. The Kings Creek catchment, which is 11.7 km^2 in area, is in the north-west portion of the FIFE area in the Konza Prairie preserve. Fig. 2 shows the division of the catchment into subcatchments – the number ranging from 5 to 66 depending on the scale. All subcatchments represent hydrologically consistent units in that runoff flows out of the subcatchments through one flow point, and that the surface runoff flux across the other boundaries is zero.

Eq. (A2) provides the relationship between variability in topography and soil and variability in local water table depths and soil moisture. Wood et al. (1990) have shown that the variability in topography dominates variability in soil properties for Kings Creek. The TOPMODEL theory uses the topographic-soil index to predict local water fluxes and soil moisture. Further, as discussed earlier, larger catchments can be considered to be composed of a population of smaller catchments that are statistically similar but whose actual

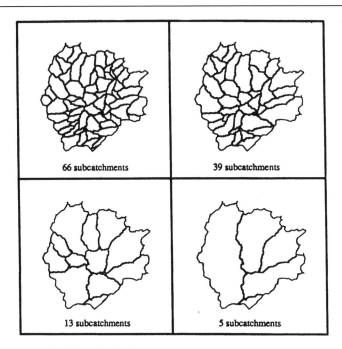

Fig. 2 Natural subcatchment divisions for Kings Creek, Kansas.

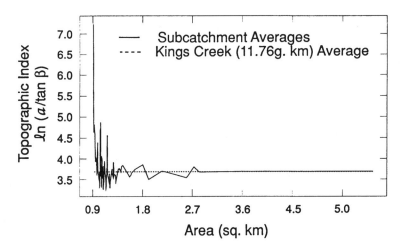

Fig. 3 Comparisons of λ, mean of ln(α/tanβ)), for the subcatchments of Kings Creek, shown in Fig. 2. Each pixel is 0.9 km².

patterns vary quite considerably. The question remains: at what catchment scale is the sample of hillslopes and small catchments sufficiently large so that their actual patterns of the soil-topographic index can be represented statistically? The average value of the topographic index, λ, was calculated for each of the subcatchments shown in Fig. 2 and plotted against subcatchment area (Fig. 3). Each pixel is 900 m². The behaviour of the catchment shows that at small scales there is extensive variability in hillslope forms leading to variability in λ, but at a scale of approximately 1 km² the increased sampling of hillslopes and small catchments leads to a decrease in the difference between topographies.

Wolock (personal communication) has found similar behaviour over a wider range of scales for Sleepers River. Fig. 4 gives his results for λ over catchment scales up to

approximately 45 km². Again, there appears to be a significant decrease in λ at about 1 to 2 km².

3.2 Scaling of vegetation

In the first experiment, scaling of the topographic index was explored due to its role in subsurface water fluxes and the redistribution of soil moisture. Vegetation type and density determine the stomatal and canopy resistances, and therefore transpiration rates in the water and energy balance models (see eqs. (A3)–(A5)). What can be said about the scaling behaviour of satellite derived estimates for vegetation?

Wood and Lakshmi (1993) used high resolution thermatic mapper (TM) satellite data to derive the normalized differ-

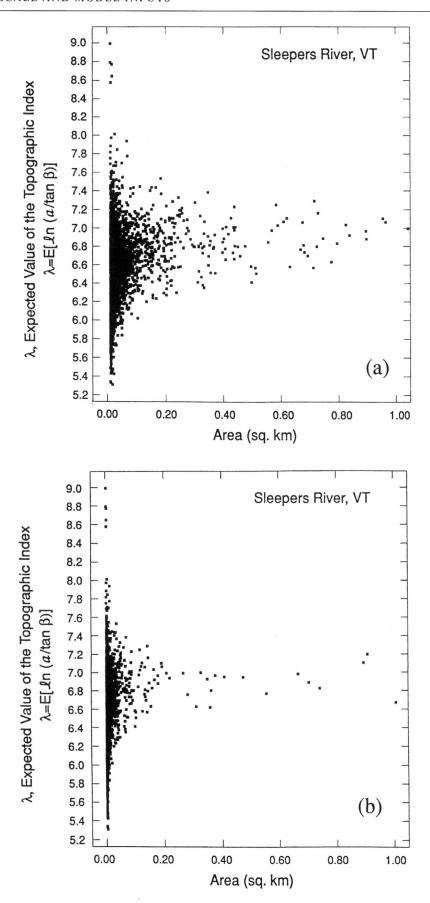

Fig. 4(a) and (b).

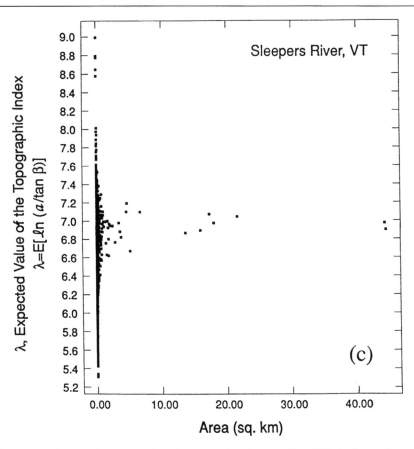

Fig. 4 Comparisons of λ, mean of $\ln(\alpha/\tan\beta)$, for the subcatchments of the Sleepers River, VT: (a) for catchments up to 1 km², (b) for catchments up to 5 km², (c) for catchments up to 45 km². (After Wolock, personal communication.)

ence vegetation index (*NDVI*), latent heat and sensible heat fluxes for the August 15, 1987 overpass and to investigate their scaling properties. The scaling for the vegetation will be reviewed here. The resolution of TM is 30 m for bands 1 through 5, and 120 m for the thermal band. The scaling question investigated here is whether averaging the TM bands prior to calculating *NDVI* provides the same derived quantities as would be found by calculating the quantities at the TM resolution and averaging. The equivalence of the two approaches depends on the degree of non-linearity represented by functions that relate *NDVI* to TM data.

The following procedure was used. The *NDVI* was calculated at the 30 m TM resolution using:

$$NDVI = \frac{(B_4 - B_3)}{(B_4 + B_3)} \qquad (2)$$

where B_3 represents band 3 (0.63–0.69 μm) and B_4 represents band 4 (0.76–0.90 μm). The first often being referred to as the red and the latter as the near infrared bands. The *NDVI* image corresponding to a TM scene acquired over the FIFE area for August 15, 1987, is given in Fig. 5. The TM scene was fully calibrated before the calculations were carried out.

For the aggregated scales, two procedures were followed. One was to aggregate spatially the TM bands and then use eq. (2), while the second procedure is to aggregate spatially the *NDVI* based on the 30 m TM data. This procedure was used for aggregation levels of 300 m × 300 m, 750 m × 750 m and 1500 m × 1500 m. A resolution equivalent to AVHRR would lie between the last two cases. Fig. 6 shows the aggregated *NDVI*, using the second procedure, for the aggregation level of 300 m × 300 m. Comparisons between the two aggregation procedures can be best shown by a scatter plot between the aggregated 30-m-based *NDVI* and the *NDVI* derived using aggregated TM bands; these comparisons are presented in Fig. 7.

One striking observation arises from comparing Figs. 5–7. Notice that the detailed structure observable in Fig. 5 is lost in Fig. 6, and yet the averaged *NDVI*s from the two aggregation schemes are essentially the same as can be seen in the scatter plot of Fig. 7. Fig. 7 does show that a small bias exists between the two aggregation procedures but its magnitude is rather insignificant. These results indicate that *NDVI* calculated from spatially averaged TM (or lower resolution AVHRR data) will be equivalent to the *NDVI* scaled up from the full resolution image.

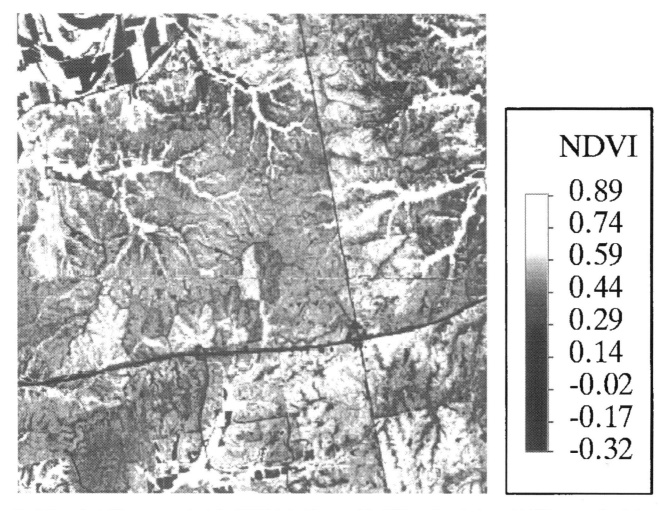

Fig. 5 Normalized difference vegetation index (*NDVI*) derived for part of the FIFE area from the August 15, 1987 overpass. Resolution is 30 m.

4. CHANGING SCALE AND DERIVED HYDROLOGICAL RESPONSES

4.1 Macroscale modeling

In a manner similar to the investigation of the scaling properties in topography, the scaling in infiltration and evapotranspiration were also investigated. For this study the water balance model described in Famiglietti et al. (1992) (see Appendix A) was applied to the Kings Creek catchment of the FIFE area in Kansas. For a rainfall storm on August 4, 1987, the average runoff for the subcatchments shown in Fig. 2 was calculated two times and plotted in Fig. 8 against subcatchment area measured in pixels. Notice that the runoff, Q_t is normalized by the average precipitation, $\bar{P}$. The same type of plot was done for selected times during an interstorm period that extended from July 18 through July 31, 1987 and is presented as Fig. 9. The behaviour of the catchment shows that at small scales there is extensive variability in both storm response and evaporation. This

variability appears to be controlled by variability in soils and topography whose length scales are of the order of 10^2–10^3 m – the typical scale of a hillslope. With increased scale, the increased sampling of hillslopes leads to a decrease in the difference between subcatchment responses.

These results are not too surprising given the linkage within the model between topography and the water balance fluxes – namely that variations in topography play a significant role in the spatial variation of soil moisture within a catchment, setting up spatially variable initial conditions for both runoff from rainstorms and evaporation during interstorm periods.

The results also suggest that at larger scales it would be possible to model the responses using a simplified macroscale model (given in Appendix A as (A6) and (A7)) based on the statistical representation of the heterogeneities in topography, soils and atmospheric forcings (rainfall and potential evaporation). Predictions based on these equations are also shown in Figs. 8 and 9 as the 'macroscale model'. Since the

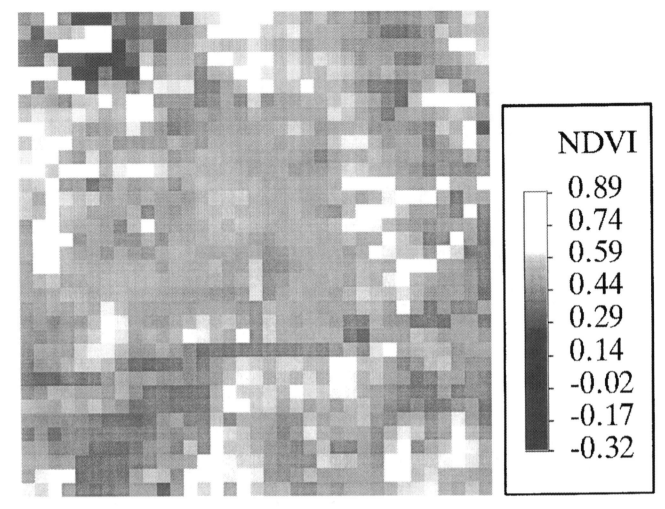

Fig. 6 The 300 m × 300 m aggregated normalized difference vegetation index (*NDVI*) for part of the FIFE area for August 15, 1987. The image was derived using data from Fig. 5.

macroscale model is scale invariant, it appears as a straight line in these figures.

4.2 Scaling remotely sensed soil moisture

To date only a very limited number of catchments have been analyzed in the manner described here. Furthermore, they have all had moderate relief and are located in regions with humid climates. For these, the REA-scale appears to be quite consistent at about 1–2 km² for both the runoff and evaporation processes. Clearly additional catchments representing a broader range of climates and catchment sizes need to be analyzed before definitive statements concerning the REA-scale can be made.

To investigate whether these scaling results are model determined or reflective of actual hydrological processes, a similar analysis was done using airborne data from the MAC-HYDRO field experiment of 1990 in Mahantango Creek, PA, a USDA experimental catchment. This experiment focused on estimating soil moisture through passive

microwave (L-band) radiation using the PBMR sensor with an effective spatial resolution of approximately 90 m and through an active radar sensor (AIRSAR) at C-, L- and P-band at a 6 m × 12 m pixel resolution. The AIRSAR remote sensing of soil moisture for MAC-HYDRO is described in Wood et al. (1993) and Lin et al. (1994) but basically the return from the radar is affected by surface soil moisture conditions. Confounding effects are due to topography, roughness and vegetation – especially large forested areas which have high reflectivity.

Much of the catchment is covered with pasture and small grains and the return in the L-band provides a good estimate of the surface soil moisture. The catchment was divided into 19 sub-catchments that ranged in size up to 3.5 km². The division was done in a manner similar to Kings Creek which is shown in Fig. 2. Fig. 10 plots the average return with catchment scale. Due to the small size of Mahantango Creek and the large areas of forest, the variance hasn't settled down as fast as that shown for the modeled results in FIFE. Nonetheless, the same behaviour can be observed, again in

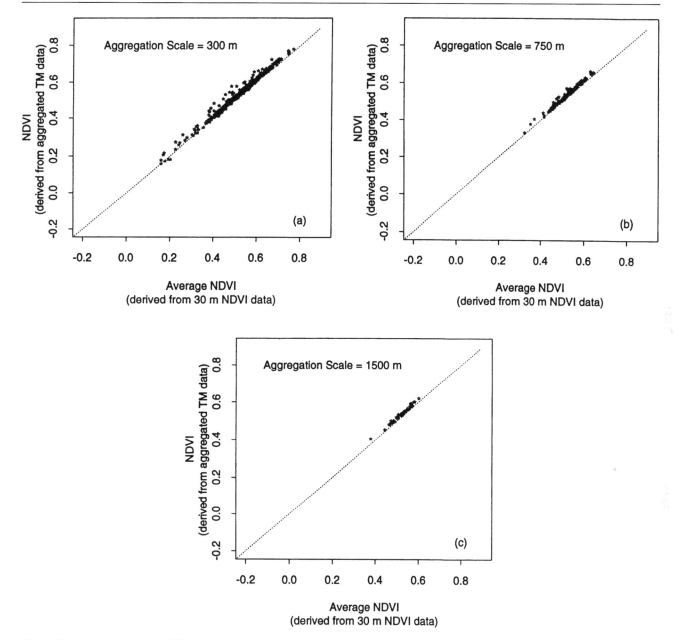

Fig. 7 Comparisons between *NDVI* derived from aggregated *NDVI* data of Fig. 5 and derived from eq. (2) using aggregated thermatic mapper (TM) data. Levels of aggregation are (a) 300 m × 300 m, (b) 750 m × 750 m, (c) 1500 m × 1500 m.

the range of 1–2 km² – our proposed REA-scale. The importance of the AIRSAR remote sensing results is that it provides an independent assessment based on measurements of the scaling behaviour of soil moisture.

5. LUMPED VERSUS DISTRIBUTED MODELS

Fig. 1 presents a framework for considering the relationship between distributed and lumped models. In an earlier section, the behaviour of aggregated inputs and hydrological responses lead to the concept of the *representative elementary*

area, a scale where a statistical representation can replace actual patterns of variability. In this section we compare the output between a macroscale, distributed model and a lumped model.

The macroscale model is based on the model described as 'model-b' in Appendix A. This model has been applied to the intensive field campaign periods (IFCs) during FIFE of 1987 and can include variability in topography, soils, net radiation and vegetation. The first two, topography and soils, lead to variations in soil moisture under the TOPMODEL framework; the latter two lead to variations in potential and actual transpiration.

A *lumped* representation (or what will also be referred to as

(a)

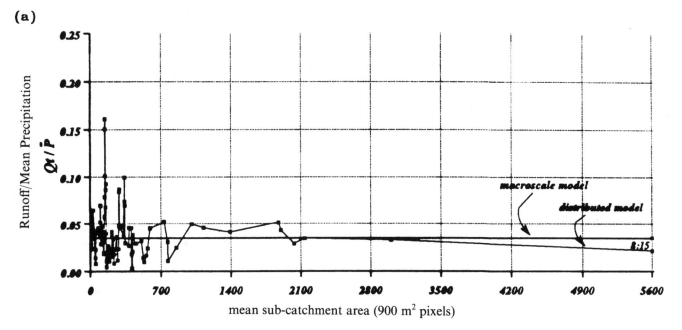

(b)

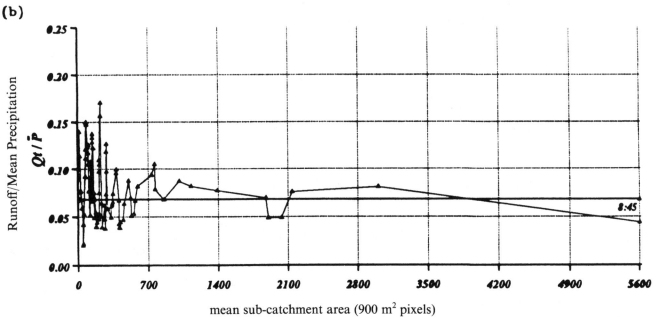

Fig. 8 Comparison of storm runoff generated from the distributed model-a (see Appendix A) and from the macroscale water balance model (eq. (A6)) for two time intervals on August 4, 1987; (a) 8:45 a.m. and (b) 9:30 a.m.

a *one-dimensional* representation) is obtained by using spatially constant values for all of the above variables. The effect of representing the distributed model by a lumped model, or equivalently by replacing the spatially variable parameters and inputs by average values, will depend on nonlinearities in the model. Conceptually this can be seen by considering a second order Taylor's series expansion about the mean for the function $y = g[x, \theta]$ where θ are fixed parameters and x variable with mean $\mu(x)$ and variance $\sigma(x)$. A first order approximation for y is $\mu_1(y) \approx g[\mu(x), \theta]$, while a second order approximation would be

$$\mu_2(y) \approx g[\mu(x), \theta] + \frac{1}{2} \frac{d^2 g}{dx_2}\bigg|_{\mu(x)} \sigma(x) \tag{3}$$

Differences between $\mu_1(y)$ and $\mu_2(y)$ depend on the magnitude of the second term in eq. (3) – the sensitivity term. As an illustrative example, consider the estimation of downslope subsurface flows, q_i, within TOPMODEL with and without considering variability in the local water table z_i. TOPMODEL relates q_i to z_i by $q_i = T_i \tan\beta \exp(-f z_i)$. Thus a first order approximation of the mean subsurface flow would be

$$\mu_1(q_i) = T_i \tan\beta \exp(-f\bar{z}) \tag{4}$$

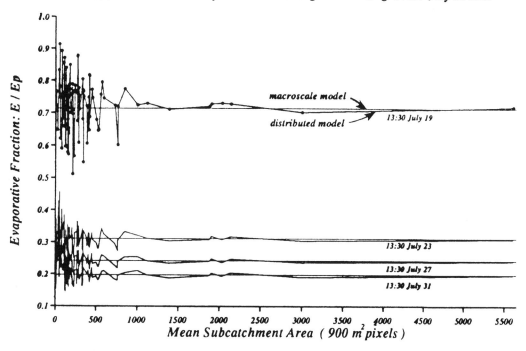

Fig. 9 Comparison of interstorm evapotranspiration from the distributed model-a (see Appendix A) and from the macroscale water balance model (eq. (A7)) for four times during the July 18–31, 1987 interstorm period.

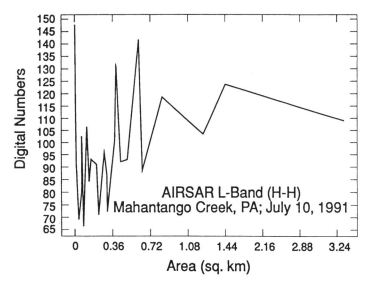

Fig. 10 Comparisons of the average radar return (L-band, HH-polarization) in digital numbers for subcatchments within Mahantango Creek, PA. Data from the July 10, 1990 AIRSAR aquisition.

while a second order approximation would be

$$\mu_2(q_i) = T_i \tan\beta \exp(-f\bar{z}) + \tfrac{1}{2}\{T_i \tan\beta f\}^2 \exp(-\bar{f}\bar{z})\sigma(z_i) \quad (5)$$

If we scale $\mu_2(q_i)$ by $\mu_i(q_i)$ and use eq. (A2) to recognize that

$$\sigma(z_i) = f^2 \sigma\left(\ln\frac{aT_e}{T_i \tan\beta}\right) \quad (6)$$

we obtain

$$\frac{\mu_2(q_i)}{\mu_1(q_i)} = 1 + 0.5 \sigma\left(\ln\frac{aT_e}{T_i \tan\beta}\right) \quad (7)$$

Analysis of the soil-topographic index for Kings Creek yields a variance of 3.25. This results in the first order estimate for q_i being biased low by approximately 65%. Since the subsurface flows and the local water table are related and since the

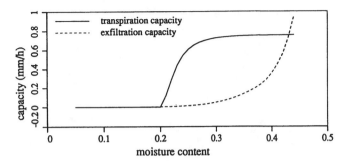

Fig. 11 Transpiration and soil exfiltration capacities versus volumetric soil moisture content (after Famiglietti and Wood, 1992).

local water table depth affects the surface soil moisture which subsequently determines the soil evaporation and infiltration rates, it's clear that the lumped model may very well lead to significant biases in the water balance fluxes.

For more complex models the sensitivities must be determined through simulation. For certain functions the sensitivities will change with the state of the catchment (wet or dry). For example Fig. 11 gives the vegetation transpiration and soil exfiltration capacities used to model the FIFE data (Famiglietti and Wood, 1992). Notice that at low and high soil moisture values the transpiration capacity function is essentially linear and the sensitivity would be low to soil moisture variations in these ranges. For volumetric moisture contents in the range 0.2–0.3, the sensitivity of the transpiration capacity function is high. As can be seen from Fig. 11, sensitivity characteristics for soil exfiltration capacity would be high for soil moisture values greater than about 0.3.

To test the sensitivity due to dry soil conditions and to compare the distributed water-energy balance model to a lumped representation (one-dimensional model or a first order model), comparisons were made between the models for 5 days during the October 1987 FIFE intensive field campaign, IFC-4. This period had the driest conditions observed during the 1987 experiment. Fig. 12 shows the simulations for October 5–9, 1987. The models were run at a 0.5 hour time step to capture the diurnal cycle in potential evapotranspiration. Three models are compared: a fully distributed model, a macroscale model in which the spatial variability is considered statistically and a lumped one-dimensional model in which parameters and inputs are spatially constant.

The one-dimensional model predicts well the evapotranspiration during the morning and late afternoon when the atmospheric demand is low, but fails to predict this flux accurately during the middle portion of the day when soil and vegetation controls limit the actual evapotranspiration. It is during this period that the sensitivity is high and by ignoring the spatial variability in soil moisture the lumped model severely underestimates the catchment-scale evapotranspiration. During wet periods, the one-dimensional model may work quite well. This complicates the linkage between a

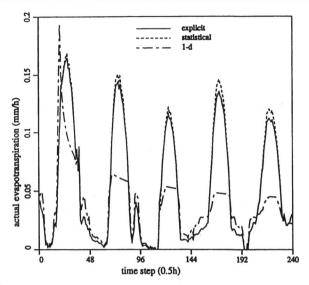

Fig. 12 Computed catchment-average evapotranspiration from the distributed model-b (see Appendix A), the statistically aggregated macroscale model, and a lumped, one-dimensional model with spatially constant parameters. Results are for Kings Creek, for October 5–9, 1987.

distributed and lumped representation since the appropriateness of the simpler representation varies with the state of the system.

6. RESULTS AND DISCUSSION

The purpose of the paper is to review recent results for the scaling of water and energy fluxes from the land component of the climate system. Three sets of experiments were presented.

The first was the aggregation of distributed inputs to determine their scaling properties and to determine whether a statistical respresentation for these parameters could be used. For topography, it appears that for catchment scales larger than about 1–2 km^2, a statistical representation is reasonable. The second part of this experiment studied scaling of the normalized difference vegetation index (*NDVI*) as derived from a thermatic mapper (TM) overpass of the FIFE area on August 15, 1987. Variations in surface conditions due to vegetation characteristics as well as topography and soils lead to significant variation in the TM-derived variables, as is shown in the presented images. Nonetheless, aggregated values of the TM band data gave accurate estimates of the aggregated *NDVI* derived from the 30 m TM data.

The second set of experiments analyzed the hydrological response at the catchment scale (but could easily be at a GCM grid scale) in which spatial variability in topography, soils and hydrological inputs (rainfall, in this case) resulted in spatially variable responses. These results support the concept of the representative elementary area (REA) (Wood et al., 1988) and its usefulness in determining the scale at which the macroscale model is a valid model for the scaled process. The results of the experiments carried out here suggest that the REA concept has wide applicability for a range of climate problems and that it appears that the REA will be of the order of a few (1.5–3) correlation lengths of the dominant heterogeneity. At scales larger than the REA scale, there has been enough 'sampling' of the heterogeneities that the average response is well represented by a macroscale model with average parameters.

The third experiment compared evapotranspiration derived from distributed models with that derived from a lumped model. The models simulated five dry days during IFC-4 of the FIFE 1987 experiment. The nonlinear behaviour of the soil and vegetation control of evapotranspiration (with respect to soil moisture) coupled the dry conditions and high mid-day potential evapotranspiration, resulted in the lumped model in underestimation of the evaporative fluxes. These results wouldn't be observed for very wet or very dry conditions, showing the subtle difficulties in understanding whether models can be represented by averaged parameters and inputs.

Current research suggests two competing approaches for handling subgrid heterogeneity:

(1) The first approach is based on the belief that subgrid processes have significant effect on processes at GCM-scales and that the nonlinearity in subgrid scale processes prevents simple scaling.

(2) The second approach is to ignore the variability in subgrid processes, and represent these processes at larger scales through models with effective parameters. This is essentially the approach of the constant canopy biospheric models where horizontal variability is ignored. It is also the approach of using small-scale micrometeorological field studies for calibration (Sellers and Dorman, 1987; Sellers et al., 1989).

The results from the experiments presented here show a rather more complicated picture. One in which macroscale models can be constructed that account for observed variability across catchments without having to account for the actual patterns of variability. Experiments to date suggest that these macroscale models will accurately predict water and energy fluxes over a wide range of catchment conditions. With regards to one-dimensional or lumped models, they may or they may not work depending on whether the catchment conditions (soil moisture levels, potential evapotranspiration, etc) lead to significant nonlinearities.

The results presented in this paper must be balanced with the knowledge that the presented experiments were neither exhaustive nor complete. For example, the satellite experiments represented a particular condition in which the range of temperatures was reasonably small, resulting in effectively linear models that transfer radiances to fluxes. Whether such ranges are typical of natural systems is unknown until a greater number of analyses are done.

It is hoped that the experiments presented in this paper motivate related research by providing a wider range of climatic data that can help resolve the basic issue concerning scaling in natural systems. What must be determined are the scaling properties for reasonably sized domains in natural systems where the range of variability (in vegetation, rainfall, radiance, topography, soils, etc) is reflective of these natural systems.

ACKNOWLEDGEMENTS

The research presented here was supported in part by the National Aeronautics and Space Administration (NASA) through Grants NAGW-1392 and NAG-5-899; this support is gratefully acknowledged. The author also wish to thank Jay Famiglietti who helped develop the water and energy balance model for Kings Creek, Dom Thongs for helping with running the model and with the graphics and V. Lakshmi who analyzed the TM data. Finally, thanks to Keith Beven, M. Sivapalan and Jay Famiglietti whose extensive discussions over the last five years have helped form my ideas on aggregation and scaling.

REFERENCES

ABBOTT, M.B., J.C. BATHURST, J.A. CUNGE, P.E. O'CONNELL, and J. RASMUSSEN, 'An Introduction to the European Hydrological System-Système Hydrologique Europeen SHE, 1, History and Philosophy of a Physically-Based, Distributed Modelling System', Journal of Hydrology, 87, 45–59, 1986a.

ABBOTT, M.B., J.C. BATHURST, J.A. CUNGE, P.E. O'CONNELL, and J. RASMUSSEN, 'An Introduction to the European Hydrological System-Système Hydrologique Europeen SHE, 2, Structure of a Physically-Based, Distributed Modelling System', Journal of Hydrology, 87, 61–77, 1986b.

BEVEN, K.J., 'Hillslope Runoff Processes and Flood Frequency Characteristics', in A.D. Abrahams (ed.), Hillslope Processes, Allen and Unwin, Boston, 1986, pp 187–202.

BEVEN, K.J., 'Scale Considerations', in D.S. Bowles and P.E. O'Connell (eds.), Recent Advances in the Modeling of Hydrologic Systems, Kluwer Academic Publishers, Dordrecht, 1988, pp 357–372.

BEVEN, K.J. and M.J. KIRKBY, 'A Physically Based, Variable Contributing Area Model of Basin Hydrology, Hydrol. Sci. Bull., 24(1), 43–69, 1979.

BEVEN, K.J., E.F. WOOD and M. SIVAPALAN 'On Hydrological Heterogeneity: Catchment Morphology and Catchment Response', Journal of Hydrology, 100, 353–375, 1988.

BINLEY, A.M., J. ELGY and K.J. BEVEN, 'A Physically-Based Model of Heterogeneous Hillslopes', Water Resources Research, 25, 1219–1226, 1989.

BROOKS, R.H. and A.T. COREY, 'Hydraulic Properties of Porous Media', Hydrology Paper No. 3, Colorado State University, Ft. Collins, Colorado, 1964.

BRUTSAERT, W., Evaporation into the Atmosphere: Theory, History, and Applications, D. Reidel Publishing Company, Dordrecht, 1982.

CHARNEY, J., W. QUIRK, S. CHOW and J. KORNFIELD, 'A Comparative Study of the Effects of Albedo Change on Drought in

Semi-Arid Regions', Journal of the Atmospheric Sciences, 34, 1366–1385, 1977.

DICKINSON, R.E., 'Modeling Evapotranspiration in Three Dimensional Global Climate Models' in Climate Processes and Climate Sensitivity, Geophysical Monograph 29, Maurice Ewing Volume 5, American Geophysical Union, Washington, D.C., 58–72, 1984.

EAGLESON, P.S., 'Climate, Soil and Vegetation', Water Resources Research, 14(5), 705–776, October, 1978.

FAMIGLIETTI, J.S., 'Aggregation and Scaling of Spatially-Variable Hydrological Processes: Local, Catchment-Scale and Macroscale Models of Water and Energy Balance', Ph.D. Dissertation, Department of Civil Engineering and Operations Research, Princeton University, October, 1992, pp 203.

FAMIGLIETTI, J.S. and E.F. WOOD, 'Multi-Scale Modeling of Spatially-Variable Water and Energy Balance Processes', Water Resources Research, 30(11), 3061–3078, November, 1994a.

FAMIGLIETTI, J.S. and E.F. WOOD, 'Application of Multi-Scale Water and Energy Balance Model on a Tallgrass Prairie', Water Resources Research, 30(11), 3079–3094, November, 1994b.

FAMIGLIETTI, J.S. and E.F. WOOD, 'Effects of Spatial Variability and Scale on Areal-Average Evapotranspiration', Water Resources Research, 31(3), 699–712, March 1995.

FAMIGLIETTI, J.S., E.F. WOOD, M. SIVAPALAN and D.J. THONGS, 'A Catchment Scale Water Balance Model for FIFE', Journal of Geophysical Research, 97(D17), 18997–19007, November, 1992a.

LIN, D.-S., E.F. WOOD, M. MANCINI, P. TROCH and T.J. JACKSON, 'Comparisons of Remotely Sensed and Model Simulated Soil Moisture over a Heterogeneous Watershed', Remote Sensing of Environment, 48, 159–171, 1994.

LIN, D.-S., E.F. WOOD, K.J. BEVEN and M. MANCINI, 'Soil Moisture Estimation During Mac-Europe'91 Using AIRSAR', 25th International Symposium on Remote Sensing and Global Environmental Change, Graz, Austria, April, 1993.

MOORE, I.D., E.M. O'LAUGHLIN and G.J. BURCH, 'A Contour-Based Topographic Model for Hydrological and Ecological Applications', Earth Surface Processes Landforms, 13, 305–320, 1988.

PANICONI, C. and E.F. WOOD, 'A Detailed Model for Simulation of Catchment Scale Subsurface Hydrologic Processes', Water Resources Research, 29(6), 1601–1620, 1993.

SELLERS, P.J., and J.L. DORMAN, 'Testing the Simple Biosphere Model (SiB) Using Point Micrometeorological and Biophysical Data', Journal of Climate and Applied Meteorology, 26, 622–651, 1987.

SELLERS, P.J., Y. MINTZ, Y.C. SUD and A. DALCHER, 'A Simple Biosphere Model (SiB) for use within General Circulation Models', Journal of the Atmospheric Sciences, 43, No. 6, 1986.

SELLERS, P.J., F.G. HALL, G. ASRAR, D.E. STREBEL and R.E. MURPHY, 'The First ISLSCP Field Experiment (FIFE)', Bulletin of the American Meteorological Society, 69, 22–27, 1988.

SELLERS, P.J., W. J. SHUTTLEWORTH, J.L. DORMAN, A. DALCHER and J.M. ROBERTS, 'Calibrating the Simple Biosphere Model for Amazonian Tropical Forest using Field and Remote Sensing Data. Part I: Average Calibration with Field Data', Journal of Applied Meteorology, 28(8), 727–759, 1989.

SHUKLA, J. and Y. MINTZ, 'The Influence of Land-Surface Evapotranspiration on Earth's Climate', Science, 215, 1498–1501, 1982.

SIVAPALAN, M., K.J. BEVEN and E.F. WOOD, 'On Hydrological Similarity: 2. A Scaled Model of Storm Runoff Production', Water Resources Research, 23, 2266–2278, l987.

SMITH, R.E. and R.H.B. HEBBERT, 'A Monte Carlo Analysis of the Hydrologic Effects of Spatial Variability of Infiltration', Water Resources Research, 15, 419–429, 1979.

SUD, Y. C., P. SELLERS, M. D. CHOW, G. K. WALKER and W. E. SMITH 'Influence of Biosphere on the Global Circulation and Hydrologic Cycle – A GCM Simulation Experiment', Agricultural and Forestry Meteorology, 52, 133–188, 1990.

TROCH, P.A., M. MANCINI, C. PANICONI and E.F. WOOD, 'Evaluation of a Distributed Catchment Scale Water Balance Model', Water Resources Research, 29(6), 1805–1818, 1993.

WALKER, J.M. and P.R. ROWNTREE, 'The Effect of Soil Moisture on Circulation and Rainfall in a Tropical Model', Quart. J.R. Meteor. Soc., 103, 29–46, 1977.

WOOD, E.F. and V. LAKSHMI, 'Scaling Water and Energy Fluxes in Climate Systems: Three Land-Atmospheric Modeling Experiments', Journal of Climate, 6(5), 839–857, 1993.

WOOD, E.F., M. SIVAPALAN and K. BEVEN 'Scale Effects on Infiltration and Runoff Production', in Conjunctive Water Use, S.M. Gorelick (ed.), IAHS Publ. 156, 375–390, 1986.

WOOD, E.F., K.J. BEVEN, M. SIVAPALAN and L. BAND, 'Effects of Spatial Variability and Scale with Implication to Hydrologic Modeling', Journal of Hydrology, 102, 29–47, 1988.

WOOD, E.F., M. SIVAPALAN and K.J. BEVEN, 'Similarity and Scale in Catchment Storm Response', Reviews in Geophysics, 28(1), 1–18, February 1990.

WOOD, E.F., D.-S. LIN, M. MANCINI, D. THONGS, P.A. TROCH, T.J. JACKSON and E.T. ENGMAN, 'Intercomparisons Between Passive and Active Microwave Remote Sensing, and Hydrological Modeling for Soil Moisture', Advances in Space Research, 13(5), 167–176, 1993.

WOOD, E.F., D.P. LETTENMAIER and V.G. ZARTARIAN, 'A Land Surface Hydrology Parameterization with Sub-Grid Variability for General Circulation Models', Journal of Geophysical Research, 97/D3, 2717–2728, February 28, 1992.

Appendix A: Spatially-distributed water and energy balance models

As shown by Beven and Kirkby (1979), variations in topography play a significant role in the spatial variation of soil moisture within a catchment, setting up spatially variable initial conditions for both runoff from rainstorms and evaporation during interstorm dry periods. Beven and Kirkby (1979) were the first to develop a saturated storm response model (TOPMODEL). This model has been further expanded to include infiltration excess runoff (see Beven, 1986; Sivapalan et al., 1987), interstorm evaporation (Famiglietti et al., 1992) and a coupled water and energy balance model (Famiglietti and Wood, 1994a). These latter two models will be described below.

1. Grid element fluxes

At the surface of each grid element, the coupled water-energy balance model (Famiglietti and Wood, 1994a) (which will be referred to as model-b) recognizes bare and vegetated land cover. Vegetation is further partitioned into wet and dry canopy. The soil column between the land surface and the water table is partitioned into a near surface root zone and a deeper transmission or percolation zone. At each grid element in the catchment, a land surface energy balance is used to calculate the potential evaporation for bare soil, unstressed transpiration for the dry canopy, and evaporation from the wet canopy. A canopy water balance is used to calculate the net precipitation. These variables, in conjunction with precipitation on bare soils, constitute the *atmospheric forcing* in the model.

The earlier water balance model of Famiglietti et al. (1992) (which we will refer to as model-a) consisted of a single soil zone, and used computed potential evapotranspiration, E_p, as the interstorm atmospheric forcing. Land cover consisted only of bare soil even though vegetated surfaces were considered implicitly through the computation of the E_p.

The storm response portion of the models captures the spatial distribution of local characteristics, such as topography and soil type, and their role in partitioning precipitation into runoff, infiltration into the unsaturated zone and percolation from the unsaturated to the saturated zone. The interstorm portion of the model determines whether atmospherically demanded evapotranspiration (potential evapotranspiration, E_p) can be met by the soil-vegetation system. At locations where it can be met, actual evapotranspiration, E, is at the potential rate, at locations where it can't be met, the actual rate is at some lower, soil or vegetation controlled rate.

2. Infiltration and runoff

2.1 SOIL DESCRIPTION

Soil type, texture, and properties are modeled using the description proposed by Brooks and Corey (1964). The five parameters utilized in this description include the saturated hydraulic conductivity, the saturation moisture content, the residual moisture content, the pore size distribution index, and the bubbling pressure, or the height of the capillary fringe above the water table. Using this soil parameterization, soil moisture and hydraulic conductivity in unsaturated soils can be described in terms of the matric head.

2.2 LOCAL COMPUTATION OF VERTICAL SOIL MOISTURE TRANSPORT

The equations for vertical transport of soil moisture for model-b include infiltration into bare and vegetated soils, evaporation from bare soil, transpiration by vegetation, capillary rise from the water table, drainage from the root zone and transmission zone, and runoff from bare and vegetated soils. Each of these

vertical moisture fluxes depends on the soil moisture status of the local root zone or the transmission zone, and the local soil properties. The infiltration, evapotranspiration and surface runoff fluxes also depend on local levels of atmospheric forcing. Canopy and soil water balance equations are applied at each grid element in the catchment to monitor the states of wetness in the local canopy, root zone, and transmission zone.

For model-a, the infiltration and evaporation processes consider only bare soil. The atmospheric forcings of precipitation and potential evapotranspiration are provided as inputs to the model. As in model-b, it is determined by the model whether the soil system can infiltrate the precipitation or provide the necessary water during evaporation to satisfy the atmospheric demand.

Infiltration is computed using the time compression approximation to Philip's equation to compute a local infiltration rate, g_i, under local time varying rainfall, p_i. The rate g_i is

$$g_i = \min \left[g_i^*(G), p_i \right] \tag{A1}$$

in which G is the cumulative infiltration during the storm and g_i^* the local infiltration capacity, which is a function of initial soil wetness, G and soil parameters. Infiltration excess direct runoff occurs when p_i exceeds g_i^*.

2.3 WATER TABLE DYNAMICS

Saturated subsurface flow between catchment elements is assumed to be controlled by the spatial variability in topographic and soil properties following the TOPMODEL approach of Beven and Kirkby (1979), Beven (1986a, b) and Sivapalan et al. (1987). This approach develops a relationship between the catchment average water table depth, $\bar{z}$, and the local water table depth, z_i, in terms of the local topographic-soil index. This relationship is

$$z_i = \bar{z} + \frac{1}{f} \left[\lambda - \ln \left(\frac{aT_e}{T_i \tan\beta} \right) \right] \tag{A2}$$

where T_i is the local soil transmissivity (saturated hydraulic conductivity divided by f), f is a parameter that describes the exponential rate of decline in soil transmissivity with depth and is assumed constant within a catchment, $\ln(T_e)$ is the areal average of $\ln(T_i)$, λ is the expected value topographic variable $\ln(a/\tan\beta)$ and is constant for a particular catchment topography, a is the area drained through the local unit contour, and β is the local slope angle.

Drainage (baseflow) between storm events is assumed to follow an exponential function of average depth to the water table (soil wetness) and has the form $Q_s = Q_o \exp(-f\bar{z})$ where $Q_o = AT_e \exp(-\lambda)$, A being the catchment area. Given a recession curve prior to a storm, Troch et al. (1993) have developed a procedure for estimating $\bar{z}$ and hence using (A2) to provide the initial patterns of local water table depths, saturated areas, and soil moisture values. The areal average water table depth is updated by consideration of catchment-scale mass balance.

2.4 EVAPOTRANSPIRATION

For model-b, evaporation from the surface is based on solving the energy balance equation, $R_n = \lambda E + H + G$, which links the energy balance to the water balance through E, the latent heat flux term. Here R_n refers to the net radiation at the land surface, H to the sensible heat flux and G the ground heat flux. A bulk transfer formulation for latent heat flux can be represented by (Brutsaert, 1982)

$$\lambda E = \frac{\rho C_p (e^*(T_l) - e_a)}{\gamma(r_a + r_{st})} \qquad (A3)$$

where ρ is the density of air, C_p is the specific heat of air at constant pressure, γ is the psychrometric constant, $e^*(T_l)$ is the saturation vapour pressure at the temperature of the surface, T_l, and e_a is the vapour pressure at a reference level above the soil or canopy surface, r_a is an aerodynamic resistance and r_{st} is a bulk stomatal resistance. Eq. (A3) can be linearized about a suitable temperature, such as the air temperature T_a, leading to the Penman-Monteith formulation.

In model-b the evaporation from the *wet canopy* is determined by the energy balance equations for the temperature of the wet vegetated surface. Setting the aerodynamic resistance consistent with the type of vegetation surface, and letting T_l represent the temperature of the wet vegetated surface yields the partitioning of R_n into λE and H. The unstressed transpiration from a canopy, E_c^*, whose density is represented by a leaf area index, LAI, is obtained from (A3) in which r_{st} is replaced with a canopy resistance $r_c = r_{st}/LAI$. Here, r_{st} is a minimum resistance corresponding to the wet vegetated surface.

The potential evaporation for *bare soil* is calculated using the nonlinear energy balance equations described above with G nonzero, r_{st} equal zero, aerodynamic resistance consistent with the particular type of soil and T_l referring to the temperature of the wet bare soil. The actual evaporation for the soil is found by applying a desorptivity based Philip-like evaporation equation like that given in (A1) for infiltration.

For a *dry canopy* the actual rate of transpiration, E_c, is related to the soil moisture through

$$\tau = \frac{\psi_s - \psi_p}{R_s + R_p} \qquad (A4)$$

where τ is the transpiration supply, ψ_s is the soil matric potential, ψ_p is the plant water potential, R_s is the hydraulic resistance of the soil and R_p is the hydraulic resistance of the plant. The actual transpiration rate is given as

$$E_c = \min[\tau, E_c^*] \qquad (A5)$$

2.5 CATCHMENT-SCALE WATER AND ENERGY FLUXES

The catchment-scale water and energy balance fluxes can be computed in two ways. The *first* is when the models are run in a 'fully distributed' mode in which the fluxes are computed grid by grid. In this mode, the grid size is usually taken to be the resolution of the digital elevation model (DEM) for the topography and therefore the resolution at which the topographic index is computed. Thus the catchment-scale water balance fluxes are just the summation over all the elements whose flux values are

determined from the process equations discussed above. In this mode, patterns of inputs (like vegetation, precipitation, radiation, etc) can be included in the flux calculations.

The *second* approach employs the similarity assumption inherent in TOPMODEL; namely that points in the catchment with the same value of the soil-topographic index respond similarly hydrologically. Since soil moisture is a dominant variable for the water and energy fluxes, this assumption appears quite reasonable. In this approach, fluxes will be determined conditional on values of the soil-topographic index, $\ln(aT_e/T_i\tan\beta)$. For cases where significant variation occurs (like vegetation characteristics) within an area, the conditioning can be taken one step further – i.e. calculate the fluxes condtional on $\ln(aT_e/T_i\tan\beta)$ and vegetation. This conditioning approach leads to macroscale models for infiltration and evapotranspiration, which are described below.

2.6 MACROSCALE MODEL FOR INFILTRATION AND RUNOFF

Using the statistical distribution of the topographic-soil index, one can determine the fraction of the catchment that will be saturated due to the local soil storage being full. These areas will generate saturation excess runoff at the rate $\bar{p}$, the mean rainfall rate. For that portion of the catchment where infiltration occurs, the local expected runoff rate at time t, m_q, can be calculated as the difference between the mean rainfall rate, $\bar{p}$, and the local expected infiltration rate, m_g. This implies that m_q and m_g are conditioned upon a topographic-soil index whose statistical distribution is central to the REA macroscale model. The difference between averaged rainfall and infiltration can be expressed as

$$m_q\{t|\ln(aT_e/T_i\tan\beta)\} = \bar{p} - m_g\{t|\ln(aT_e/T_i\tan\beta)\} \quad (A6)$$

As discussed above, m_q and m_g are time varying functions whose values at any particular time are equal for points within the catchment having the same topographic-soil index; this dependence is indicated in eq. (A6) by the |. The full development of the topographic-soil index is provided in Beven and Kirkby (1979), Beven (1986a, b), Sivapalan et al. (1987) and Wood et al. (1990). Both the local expected runoff rate and the local expected infiltration rate are (probabilistically) conditioned on the topographic-soil index, $\ln(aT_e/T_i\tan\beta)$. The runoff production from the catchment is found by integrating, usually numerically, the conditional rate over the statistical distribution of topographic-soil index.

2.7 MACROSCALE MODEL FOR EVAPOTRANSPIRATION

In a similar way, a macroscale evaporation model is developed for interstorm periods. As stated earlier, topography plays an important role in the interstorm redistribution of soil moisture and therefore in the initial conditions for the evaporation calculations. For those portions of the catchment for which the soil column can deliver water at a rate sufficient to meet the potential evapotranspiration or atmospheric demand rate, E_p, the actual rate E equals E_p; otherwise, the rate will be at a lower soil controlled rate E_s. Within the TOPMODEL framework, locations with the same value of the topographic-soil index will respond similarly; implying a macroscale model of the following form, which is conditioned on that index.

$$m_E\{t|\ln(aT_e/T_i\tan\beta)\} = \min[m_{E_s}\{t|\ln(aT_e/T_i\tan\beta)\}, \bar{E}_p(t)] \quad (A7)$$

where m_E refers to the mean evaporation rate at locations in the catchment with the same index, m_{E_s} refers to the mean soil controlled rate and $\bar{E}_p$ to the spatially average potential or atmospheric demand rate.

Scale problems in surface fluxes

J.C.I. DOOGE

Centre for Water Resources Research, University College Dublin,

Earlsfort Terrace, Dublin 2,

Ireland

ABSTRACT Partial analysis in the form of limiting analytical solutions is applied to the problem of infiltration into a semi-infinite column at constant initial moisture content. It is found as a result of such an analysis that: (a) the solutions for realistic assumptions concerning the soil moisture characteristics are bounded reasonably closely by an upper limit corresponding to constant diffusivity and ultimate infiltration rate equal to saturated conductivity and a lower limit corresponding to constant diffusivity and zero conductivity; (b) for both of the limiting cases of zero horizontal conductivity and infinite horizontal conductivity, the values of average sorptivity are insensitive to the form of the statistical distribution of spatial non-homogeneity; (c) the average sorptivity is less than the corresponding sorptivity based on the average scale parameter of a spatially variable soil and the average ultimate rate of infiltration is greater than the corresponding ultimate rate based on the average scale parameter.

1. INTRODUCTION

It is a privilege to contribute to this colloquium in honour of George Kovacs whom I valued as a colleague and a friend over a period of twenty years. In doing so I have sought to select a topic that would reflect his own special interests and his own approach to hydrologic problems. Thus, I have chosen to deal with conditions in the unsaturated zone because of his own interest in and contributions to subsurface hydrology. In dealing with this topic, I have sought for underlying simplicity because the search for the simple solution was an essential part of the approach of George Kovacs to both scientific and administrative problems.

The amount and the movement of soil moisture is a key element both in hydrological science and in applied hydrology. In relation to both the surface water phase and the groundwater phase of the hydrological cycle, a meaningful insight can be obtained and reasonable prediction can be made on the basis of simple modelling of water movement, often involving only linear models. In the case of average fluxes at the land surface, however, there are concentrated non-linearities in the form of thresholds which make such approaches based on simple areal averaging inappropriate. Soil moisture accounting and surface fluxes are of great importance in a number of areas including: (a) an adequate understanding of the hydrological cycle in a given region; (b) important areas of applied hydrology such as flood prediction and irrigation management; and (c) in the linking of hydrologic models with atmospheric models and prediction.

The present paper is concerned with the search for a simplicity in this particular area that will give us both an insight into the phenomenon concerned and a reasonable predictive capability. The emphasis is on simplicity; neither large quantities of computer output dealing with special cases nor algebraic formulations of daunting complexity can do much to promote insight into the essentials of the phenomena involved or to assist in providing a realistic solution for the problem under investigation. It is clear from the formulation of the problem of land surface fluxes that the phenomenon involves strong non-linearities. Accordingly, the equations valid at the micro-scale cannot be extended to a macro-scale problem involving spatial variability by a simple averaging of the soil parameters involved in the basic equations. The questions raised in this paper relate to a general exploration of the direction and the degree of bias in estimates of land surface fluxes estimated on the basis of such simple averaging.

The work is undertaken in the spirit of Polya (1957) who

suggested that the key to solving a complex problem may be to ignore the complexity and to solve first a related simple problem before returning with renewed insight to the original complex problem. Such an approach belongs to the tradition which has been described as partial analysis or fractional analysis and has been characterised by Kline (1965) as follows: 'Any procedure for obtaining some information about the answer to a problem in the absence of methods or time for finding a complete solution'.

The development of modern fluid mechanics owes much to such an approach through the concept of the boundary layer within which transverse acceleration was neglected but viscosity accounted for, in contrast to the main flow in which the reverse was the case (Prandtl, 1904). Perhaps, hydrology could benefit from a similar approach.

Partial analysis can evaluate the sensitivity of different types of output to simplifying assumptions. Simplified solutions for flow in the unsaturated zone that give realistic storage volumes and boundary fluxes in spite of unrealistic soil moisture profiles may be adequate for many hydrologic purposes. The present paper is an exploration of this possibility. In the exploration of the spatial variability at field scale (about 10^{-2} m), use will be made of the geometrical scaling used to parametrise from the particle scale 10^{-6} m–10^{-3} m to the pedon scale of 10^{-2} m (Miller and Miller, 1955, 1956; Kuhnel et al., 1991). Use is also made of limiting cases as providing plausible bounds on the more general but more complex model developed to give a closer point simulation of non-linear and non-homogeneous reality (Romanowicz et al., 1990). The main purpose is to explore how sensitive is the estimate of average infiltration or average exfiltration to the type and the degree of spatial variability at the field scale.

It is hoped through this approach to explore the possibility of upscaling to bridge the gap between the pedon scale of 10^{-2} m and the field scale of 10^3 m represented by the representative elementary area (Wood et al., 1988).

2. FORMULATION OF THE PROBLEM ——

2.1 Basic equations for unsaturated flow

In soil moisture accounting, we are concerned with the occurrence of water characterised by the volumetric water content $\theta(x, y, z, t)$ and the movement of water characterised by the flux vector $q(x, y, z, t)$. These two dependent variables are linked by a single linear continuity equation and by a set of three equations of motion which for realistic soil characteristics are highly non-linear. The single continuity equation relating moisture content and moisture flux is given by

$$\frac{\delta\theta}{\delta t} + \frac{\delta q_x}{\delta x} + \frac{\delta q_y}{\delta y} + \frac{\delta q_z}{\delta z} = 0 \qquad (1)$$

and the dynamic equation relating moisture flux and matric head $h(\theta)$ is given by

$$q_x = -K_x(\theta)\frac{\delta h}{\delta x} \qquad (2a)$$

$$q_y = -K_y(\theta)\frac{\delta h}{\delta y} \qquad (2b)$$

$$q_z = -K_z(\theta)\frac{\delta h}{\delta z} + K_z(\theta) \qquad (2c)$$

where the additional term in eq. (2c) represents the effect of gravity on the downward vertical flux q_z. In order to solve this set of equations it is necessary to know or assume the relationship $h(\theta)$ between the matric head h and the moisture content θ and the form of the hydraulic conductivity vector as a function of moisture content $K(\theta)$ or as a function of matric head $K(h)$.

The formulation of the above equations itself involves parametrisation from the micro-scale of a soil particle (10^{-6} m to 10^{-3} m) to the macro-scale of a soil pedon (10^{-2} m). This parametrisation involves a double integration of the Navier-Stokes equation, simplied for the case of creeping flow, over the cross-section of the individual pore space and then over the cross-section of a representative elementary volume (Dooge, 1986).

2.2 Boundary and initial conditions

To be complete, any realistic problem involving eqs. (1) and (2) requires two boundary conditions for each space dimension and one initial condition. If the area is a closed system, then the boundary conditions in the (x, y) plane will be zero flux conditions. The lower boundary condition in the z direction will be given either by the presence of a water table which necessarily involves

$$\theta = \theta_{sat} \quad \text{at} \quad z = z_o \qquad (3)$$

where z_o is the depth of the water table below the surface; or by the assumption of a zero-flux plane, i.e.

$$q_z = 0 \quad \text{at} \quad z = z_o \qquad (4)$$

where z_o is the depth of this constant flux plane below the surface. More critical in seeking an insight into soil moisture accounting are the surface boundary conditions and the initial condition.

The simplest form of surface boundary condition would be a prescribed flux or a prescribed moisture content at the surface. In the classical work on infiltration at the land surface (Philip 1954, 1957a, b, 1969), the assumption was made that the surface became saturated and began to pond

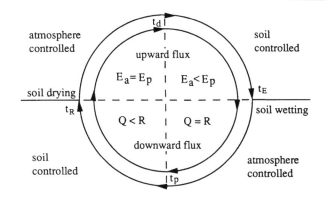

Fig. 1 Four phases of surface flux control.

immediately following the onset of rain. In the case of evaporation (i.e. exfiltration) the corresponding classical case would be instant drying of the surface following the cessation of rainfall. In both cases the simple initial condition taken was that of uniform moisture content throughout the soil profile.

In fact, there are four standard cases of surface boundary condition and these are shown on Fig. 1. If rain falls upon soil that is initially below saturation at a rate in excess of the saturated hydraulic conductivity, the surface layers of the soil will become saturated but only after a finite period of time (t_p) during which the rate of surface infiltration will be equal to the rate of precipitation. Thus, during this preponding phase the rate of downward flux at the surface is atmosphere-controlled rather than soil-controlled. Below the surface the moisture content at any given depth will be increased above that of the initial profile; as this new profile builds up the gradient in the matric head will decrease and the amount of downward soil water flux at lower levels will increase thus acting as a negative feed back on the profile build up. Following saturation of the surface, the rate of infiltration down through the land surface will become less than the rate of precipitation and the rate of infiltration will be soil-controlled rather than atmosphere-controlled. If the rainfall continues long enough this process will continue until the whole soil profile is saturated and the rate of surface infiltration drops to the value of vertical hydraulic conductivity of the saturated soil.

During dry periods following prolonged rain there will be similar switch of control from the atmosphere to the soil. Initially the soil will be able to convey water vertically upwards to the surface in order to provide evaporation at the head rate as determined by atmospheric conditions. When the surface dries to a certain point (t_d) the soil will no longer be able to supply water at this rate and there will be a switch from atmosphere-controlled evaporation to soil-controlled evaporation during which the actual evaporation will be less

than the potential rate. If the dry period were very extensive this process would continue until the soil became air dry at greater and greater depths. There are thus four distinct phases of surface flux as shown on Fig. 1:

(1) $t < t_p$: atmosphere-controlled infiltration (downward flux = rainfall rate);
(2) $t_p < t < t_R$: soil-controlled infiltration (surface saturated);
(3) $t_R < t < t_d$: atmosphere-controlled evaporation (upward flux = potential evaporation);
(4) $t_d < t < t_E$: soil-controlled evaporation (surface dessicated).

Any realistic method of soil moisture accounting, either for the purpose of hydrologic modelling or for climate modelling, must take account of all four standard cases and of the switching in control during the transition from one phase to another. For certain simplifying assumptions about the soil moisture characteristics, analytical solutions are available for all four phases but only for the simple initial condition of constant moisture content throughout the soil profile.

If the alternating periods of precipitation and evaporation are prolonged, the initial conditions for the atmosphere-controlled phases (1) and (3) can be taken as a uniform moisture distribution. The initial conditions in the cases of phases (2) and (4) are more troublesome. During the preponding phase of infiltration a profile will develop with increasing moisture content at and below the surface while maintaining the initial constant moisture content at large depths. This profile will continue to develop until the moisture content at the surface reaches saturation. It is the resulting profile of non-uniform moisture content that constitutes the initial condition for the following soil-controlled phase in which the upper boundary condition is the fixed concentration of saturation. Similarly during dry periods when the surface becomes desiccated a profile will develop in which the soil is air dry at the surface and at the same moisture content as the initial condition only at large depths. This in turn is the appropriate initial condition for the fourth phase of soil-controlled exfiltration (evaporation).

One-dimensional analytical solutions of the problem are only available for the case of uniform initial conditions.

2.3 Strategy for sensitivity search

In accordance with the partial analysis approach mentioned in the introduction, the purpose is to seek simplifications of the problem which will produce widely applicable procedures that will not involve undue errors of prediction. This involves distinguishing between those factors to which the problem is most sensitive and the remaining factors to which the result is relatively insensitive so that simplifying assumptions can be made with impunity.

An obvious candidate for investigation here is the sensitivity of the rate of soil-controlled infiltration or soil-controlled exfiltration to the soil moisture characteristics. One approach is to see if we can discover simple analytical solutions which can provide relatively narrow bounds on the more complex solutions arising from more realistic but more complex assumptions in regard to the soil characteristics. The basis of this strategy is to seek a parametrisation that would be valid for an area of uniform soil type and topography leaving a further step of parametrisation from this level to the GCM grid scale of 10^5 m through a semi-distributed model (Beven & Kirkby, 1979; Wood, this volume) or through inversion of satellite data (Feddes, this volume). This exploratory analysis is made in the present paper on the basis of a semi-infinite soil column, i.e. with the lower boundary condition at an infinite depth below the surface. The solution for a finite depth of soil column involves one or more convergent infinite series of which the first term represents the solution for the semi-infinite column. In a first attack on the problem, it would appear legitimate to proceed on the basis of the semi-infinite case. This is done to avoid additional complexity in the algebra and is unlikely to impair the result. Such an exploration is described in Section 3.1 below.

At first sight the restriction of the analytical solutions to a uniform moisture content in the profile as an initial condition seems daunting. Early attempts at tackling the problem of infiltration and exfiltration with spatial variability resorted to the device of redistributing the moisture profile to a uniform distribution at every switch of control (e.g. Milly and Eagleson, 1982). However, as shown in Section 3.2 below this approach can be improved by the less drastic redistribution from a flux-boundary condition profile to a concentration-boundary profile at the instant of ponding. In most cases this involves a discontinuity in the flux but maintains conservation of volume which is a vital condition in soil moisture accounting.

The final area of simplification is that of spatial variability and the associated horizontal fluxes of soil water movement. In relation to the specification of spatial variability, dimensional analysis at the micro-scale is used in order to reduce the separate spatial variability in the matric head function and in the hydraulic conductivity function to variability in a single length-scale parameter. The effect of the horizontal fluxes on the average infiltration over the non-homogeneous area is then examined by analysing the two extreme cases of zero horizontal hydraulic conductivity and infinite horizontal hydraulic conductivity to provide overall limits. The results of these studies are given in Sections 4.1 and 4.2 of the paper for the classical problem of instantaneous ponding or instantaneous desiccation and the adjustment to allow for the atmospheric-controlled phases is discussed in Section 4.3.

3. SIMPLIFYING THE PROBLEM

3.1 Classical one-dimensional equations

The reduction of the problem based on eqs. (1) and (2) subject to appropriate boundary conditions to a one-dimensional problem as a first step is a simplification well represented in the literature on flow through porous media. Analytical solutions have been presented for a semi-infinite soil column with an initial uniform moisture content for both a concentration boundary condition and a flux boundary condition at the surface and various assumptions in relation to the matric head $h(\theta)$ and the hydraulic conductivity $K(\theta)$.

Under the restriction of one-dimensional vertical flow, the new equation of continuity becomes

$$\frac{\delta\theta}{\delta t} + \frac{\delta q}{\delta z} = 0 \tag{5}$$

where $q(z, t)$ is the soil water flux taken vertically downwards. The equation of motion then becomes

$$q = -K(\theta)\frac{\delta h}{\delta z} + K(\theta) \tag{6}$$

where $K(\theta)$ is the hydraulic conductivity in a vertical direction and $h(z, t)$ is the matric head. In classifying the soil moisture characteristics in relation to analytical solutions and in seeking a solution of a problem characterised by eqs. (5) and (6), it is convenient to work in terms of either $\theta(z, t)$ or $h(z, t)$. This can be done by defining the hydraulic diffusivity as

$$D(\theta) = K(\theta)\frac{dh}{d\theta} \tag{7}$$

and writing eq. (6) as

$$q = D(\theta)\frac{\delta\theta}{\delta t} + K(\theta) \tag{8}$$

Alternatively, one can define the specific water capacity as

$$C(h) = \frac{d\theta}{dh} \tag{9}$$

which can be used in the continuity equation to give

$$C(h)\frac{\delta h}{\delta\theta} + \frac{\delta q}{\delta z} = 0 \tag{10}$$

Both of these substitutions ignore the hysteretic nature of the relationship between the soil water matric head and the moisture content and thus the parameters will only apply to either drying or wetting conditions. The three parameters $K(\theta)$, $D(\theta)$ and $C(h)$ are not independent of each other since a comparison of eqs. (7) and (9) confirms that diffusivity is equal to conductivity divided by water capacity.

Table 1. *Table of infiltration solutions for different types of boundary condition.*

Hydraulic diffusity	Hydraulic conductivity		
	Constant K	Linear K	Nonlinear K
Delta function D		Green and Ampt (1911) Mein and Larson (1973)	
Constant D	Carslaw and Jaeger (1946) Braester (1973)	Philip (1966) Braester (1973)	Philip (1974) Clothier et al. (1981)
Fujita D	Fujita (1952) Knight and Philip (1974)	Not solved Rogers et al. (1983)	Not solved Sander et al. (1988)

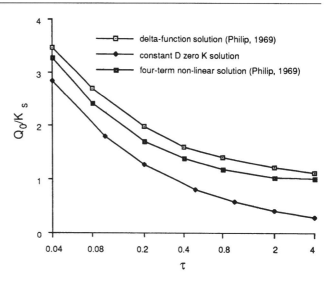

Fig. 2 Sensitivity to soil moisture characteristics.

3.2 One-dimensional analytical solutions

For reasons of space, the remaining discussion will be restricted to only two of the four phases shown in Fig. 1. It is natural when faced with this choice, to deal with the case of infiltration under instantaneous ponding and to adjust the solutions of this classical case to allow for pre-ponding conditions. The same considerations apply to the other two phases which can be handled using the solution for instantaneous desiccation and adjusting it to allow for pre-desiccation exfiltration at the head evapotranspiration rate.

Table 1 shows the closed form analytical solutions of the infiltration problem available for both types of boundary condition in respect of simplified soil moisture characteristics. The names in the table indicate the first publication of the solution for the infiltration problem, the upper name relating to the concentration boundary condition case and the lower name to the flux boundary condition case. The oldest solution due to Green and Ampt (1911) was based on an analogy with capillary tubes but was reinterpreted by Philip (1954) as corresponding to the assumption that the hydraulic diffusivity took the form of a δ-function, i.e. that it was an abrupt wetting front with no gradual diffusion below it.

A comparison of the analytical solutions for infiltration under instantaneous ponding of an originally dry soil reveals the fact that they are closely bounded by two of the simpler cases (Philip, 1969; Romanowicz et al., 1990). Philip (1969) has suggested that most of the more complex solutions can be closely approximated by

$$f(t) = \frac{S}{2t^{1/2}} + f_{ult} \qquad (11)$$

where S is termed the sorptivity.

The lower limit is provided by the case of constant D and

zero K, i.e. pure diffusion without drainage under gravity. For this case (Childs 1936), the infiltration at the surface is given by

$$f(t) = (\theta_{sat} - \theta_o)\left(\frac{D}{\pi t}\right)^{1/2} \qquad (12)$$

There does not appear to be a single upper limiting case. The chief candidate models are: (1) delta-function diffusivity with saturated conductivity above the abrupt wetting front and zero conductivity below it; and (2) constant diffusivity with a constant hydraulic conductivity for all moisture contents. For the delta-function diffusivity the solution is

$$\frac{K_{sat}t}{S_o(\theta_{sat}\theta_o)} = \frac{1}{f/K_{sat} - 1} - \log_e\left[1 + \frac{1}{f/K_{sat} - 1}\right] \qquad (13)$$

For the constant D, constant K model the solution is

$$f(t) = (\theta_{sat} - \theta_o)\left(\frac{D}{\pi t}\right)^{1/2} + K_{sat} \qquad (14)$$

For small values of time, eq. (13) gives marginally higher rates of infiltration than eq. (14) but for very large values of time, eq. (14) gives substantially higher rates. Because of the simple form of eq. (14) compared with the implicit form of eq. (13), the former case will be taken as a plausible upper bound in this first analysis. This preference is further justified by the fact that the Green-Ampt solution is not well defined for the case of exfiltration (i.e. evaporation from a bare soil surface). A comparison of the limiting solutions is shown on Fig. 2 which also shows the 4-term non-linear solution due to Philip (1969).

It can be verified that the difference between these two limiting forms is less by one or two orders of magnitude than the variability usually encountered in soil moisture charac-

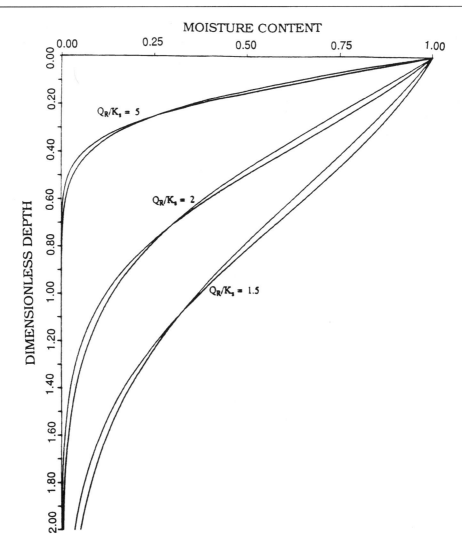

Fig. 3 Profile matching at ponding for three different ratios of precipitation Q_r to saturated conductivity K_s using the model of constant diffusivity and linear conductivity respectively.

teristics at the field scale. The problems raised by spatial variability are therefore far more important than the question of the precise choice of model in a simplified approach to dealing with the problem of spatial non-homogeneity.

3.3 Adjustment of initial conditions

For wetting or drying over a prolonged period, the restrictive assumption of constant initial moisture content does not lead to serious error but in a switching from wetting to drying and vice versa serious errors could occur. If the atmosphere-controlled phase prior to ponding in which the surface infiltration is equal to the rainfall is taken into account, then the assumption for the post-ponding conditions for constant initial moisture content is no longer valid. This can be overcome in a soil moisture accounting procedure by match-

ing the volumes of cumulative infiltration at the instant of ponding for the two cases.

For the case of the Green-Ampt model the matching is exact because a rectangular profile is assumed a priori in the model both for the constant flux boundary condition and for the constant concentration boundary condition. For other models it can be shown that, while the matching is not exact, it is remarkably close in all cases (Kuhnel et al., 1991, Dooge and Wang, 1993). Fig. 3 shows the closeness of fit for the model of constant diffusivity and linear conductivity for three different ratios of precipitation to saturated conductivity. There is a similar close matching of profile shapes at the instant of desiccation following evaporation of a saturated soil column.

The matching of the volumes of infiltration for the two types of boundary condition necessarily involves a discontinuity in the predicted rate of infiltration except in the special case of the Green-Ampt solution. However, in a soil moisture accounting procedure it is more important to maintain continuity of volume rather than continuity of flux (Wang and Dooge, 1993).

For any given simplified model, expressions are available for the time to ponding t_p (or to desiccation t_d) and for the cumulative increase V_{fp} (or decrease V_{fD}) in the volume of soil moisture in the semi-infinite soil column for the case of a constant flux of infiltration (or exfiltration) at the surface. Expressions are also available for the variation with time of the cumulative increase V_{sp} (or decrease V_{cd}) of the volume of soil moisture for the given concentration at the surface of saturation or desiccation. The reduced flux at the surface for times longer than the time to saturation (or to desiccation) can be approximated very closely by applying the concentration boundary solution starting at a time which gives the correct value at the instant of saturation (or desiccation) to estimate this declining rate of infiltration (or exfiltration).

3.4 Specification of spatial variability

In the case of a non-homogeneous area, the spatial variability of both the soil water retention function $h(\theta)$ and the hydraulic conductivity parameter $K(\theta)$ or $K(h)$ must be taken into account. The first step in simplification is to attempt to link the variation between these two soil parameters. This can be done by applying dimensional analysis on the micro-scale of the soil particle at which surface tension is the dominant force and using the result on the macro-scale appropriate to the soil pedon at which the basic equations of non-saturated flow are formulated. This was accomplished by Miller and Miller (1955, 1956) and has been proved as useful both in the laboratory and in the field.

Defining λ as any specified characteristic length, the relationship of moisture content θ to matric head h for two (or more) similiar media with identical drying and wetting contact angles must be given by a common functional F_1, i.e.

$$\theta = F_1 \left(\frac{\gamma}{\sigma} \lambda h \right) \qquad (15)$$

where γ is the weight density of water and σ is the surface tension of water in contact with the soil particle.

From a comparison of the Navier-Stokes equation at the micro-scale of the soil pore with the Darcy equation at the macro-scale of the continuum point or soil pedon, it can be deduced that the hydraulic conductivity $K(\theta)$ must be inversely proportional to the viscosity of the liquid (μ) and proportional to the square of the characteristic length λ. Thus for two similar media

$$\frac{\mu}{\gamma} \frac{K}{\lambda^2} = F_2(\theta) = F_3 \left(\frac{\gamma}{\sigma} \lambda h \right) \qquad (16)$$

From eq. (7) it can be deduced that the diffusivity D will be directly proportional to λ and from eq. (9) that specific capacity C will be inversely proportional to λ. Furthermore, it follows from eq. (12) that sorptivity S will be proportional to the square root of λ.

The variability in soil moisture characteristics in a field, which often amounts to one or two orders of magnitude, can be substantially decreased by the use of such scaling. Fig. 4 shows the striking reduction in variability in the soil moisture curve for the case of the Hupselse Beek catchment in the Netherlands (Hopmans, 1987). The wide scatter in the value of the moisture content at each standard laboratory test value of the matric head does not collapse completely to a single curve but the scatter is reduced to an amount acceptable in preliminary partial analysis. Fig. 5 shows the effect of such scale analysis on reducing the variation for seventeen points in a 1-hectare field of bare soil (Vauclin, 1984). These results indicate that it would be reasonable to base the first analysis of the problem on the specification of the variation of the linear scaling parameter λ and then to take the Miller and Miller relationships as the basis for the variation of the soil moisture head (h) and the hydraulic conductivity (K).

As already indicated in eqs. (15) and (16), the local matric head is inversely proportional to the local scaling parameter, i.e.

$$h(x, y, z, t) \sim [\lambda(x, y, z)]^{-1} \qquad (17a)$$

and the hydraulic conductivity is proportional to the square of the local scaling parameter, i.e.

$$K(x, y, z, t) \sim [\lambda(x, y, z)]^2 \qquad (17b)$$

For the hydraulic diffusivity defined by eq. (7) and the specific water capacity defined by eq. (8) the variation is directly proportional to the scaling parameter, i.e.

$$D(x, y, z, t) \sim \lambda(x, y, z) \qquad (17c)$$

$$C(x, y, z, t) \sim \lambda(x, y, z) \qquad (17d)$$

Finally, in the case of spatial variability the assumption of steady state initial condition implies a spatial variation in $(\theta_{sat} - \theta_o)$.

In exploring the limiting effects of spatial variability two simple distributions will be examined, one of them a lumped distribution and the other a distributed distribution. The simplest case of variation is where there are two different values of λ within the area. The lumped model used is that of two soil columns equal in area but with different values of the scaling factor λ. This can be considered either as a deterministic model or as a probablistic model based on an extreme form of a U-shaped distribution. In the latter case the two values of the scale factor can be expressed as

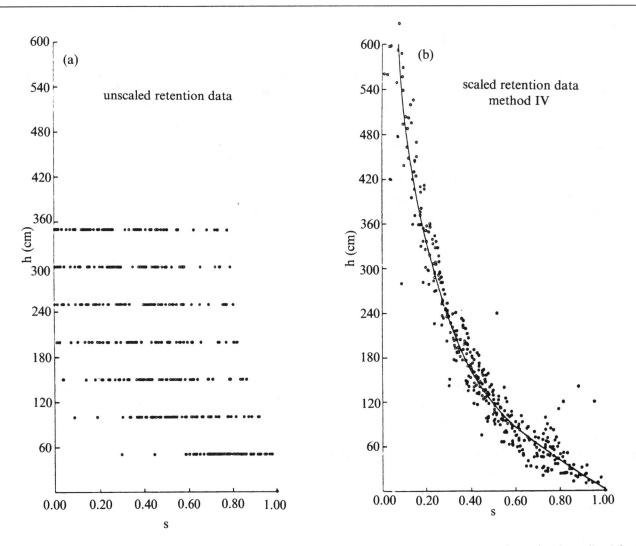

Fig. 4 Reduced variability in soil matric suction (= negative value of matric head h) versus moisture saturation ratio S by scaling (after Hopmans, 1987).

$$\lambda_1 = E(\lambda) + \sigma(\lambda)/\sqrt{2} \qquad (18a)$$

$$\lambda_2 = E(\lambda) + \sigma(\lambda)/\sqrt{2} \qquad (18b)$$

where λ is the mean value and $\sigma(\lambda)$ is the standard deviation of the scaling factor.

Field measurements indicate the distribution of the scaling factor to be unimodal with positive skew. In this treatment the distribution chosen is the gamma distribution. This is used in preference to similar skewed distributions such as the log normal because of the algebraic simplicity of the results obtained. In any case, it is seen in the course of the study that the limiting cases are not sensitive to the distribution used. If the gamma distribution is written as

$$p(\lambda) = \frac{a \exp(-a\lambda)(a\lambda)^{b-1}}{\Gamma(b)} \qquad (19a)$$

then the relative parameters of interest will be given by

$$E(\lambda) = b/a \qquad (19b)$$

$$\sigma^2(\lambda) = b/a^2 \qquad (19c)$$

$$CV(\lambda) = 1/b^{1/2} \qquad (19d)$$

The approach used can be readily adapted for other types of distribution at the cost of more complicated algebra.

4. AVERAGE FLUX FOR NON-HOMOGENEOUS AREAS

4.1 Limiting case of low horizontal conductivity

The classical studies reported in the literature on unsaturated flow in porous media were devoted to the study of the one-dimensional problem. One crude approach to the study of spatial variability in surface fluxes is to estimate the surface flux for the individual soil columns on the basis of these one-dimensional solutions and then to average the flux over the area of interest. This is equivalent to assuming that there is no interaction between the soil columns, i.e. that the horizontal

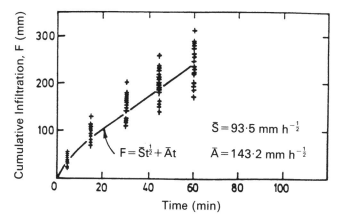

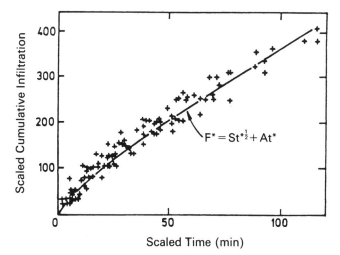

Fig. 5 Reduced variability in infiltration by scale analysis (after Vauclin, 1984).

conductivity $K_{hor} = 0$. An alternative approach is to seek a solution of the other limiting case in which there is complete mixing between the columns, i.e. $K_{hor} = \infty$. The results for finite values of K_{hor} would be expected to be intermediate between these two limiting solutions.

On the basis of the results in the literature and on the discussion of limiting cases in Section 3.1 above, the partial analysis will be made on the basis that the infiltration at the pedon scale can be described with acceptable accuracy by

$$f(t,\lambda) = \tfrac{1}{2} S(\lambda) t^{1/2} + f_{ult}(\lambda) \tag{20a}$$

where

$$S(t,\lambda) = \theta_{sat} - \theta_o)_\lambda \left(\frac{d(\lambda)}{\pi t} \right)^{1/2} \tag{20b}$$

and $D(\lambda)$ is the local diffusivity and f_{ult} is the final rate of local surface infiltration (Philip, 1969). Since eqs. (20) correspond to constant diffusivity and constant conductivity we can deduce from the resulting

$$D(\theta_{sat} - \theta_o) = Kh \tag{21a}$$

that for an initial condition of constant h_o we would have

$$(\theta_{sat} - \theta_o)_\lambda - \lambda(x, y, z) \tag{21b}$$

For this formulation, the problem of estimating the average infiltration at the field scale reduces itself to the problem of estimating the distribution of $(\lambda)^{3/2}$ and the distribution of $(\lambda)^2$. For the case of zero initial moisture control, the key problem is to express the mean and the variance of the two field parameters $S(\lambda)$ and $f_{ult}(\lambda)$ in terms of the characteristics of the distribution of λ at the pedon scale.

The result for the case of the ultimate rate of infiltration is simple and is found to be distribution-free in the sense that it depends only on the mean and the variance of the distribution of the scaling parameter (λ). For any distribution the expected value of $(\lambda)^2$ is given by

$$E[\lambda^2] = \sigma^2(\lambda) + [E(\lambda)]^2 \tag{22}$$

so that any distribution with two or more parameters which reproduces the mean and the variance of the scaling parameter in the field will give a correct value for the average value of the ultimate rate of infiltration given by

$$E[f_{ult}(\lambda)] = [1 + \{CV(\lambda)\}^2] f_{ult}[E(\lambda)] \tag{23}$$

so that the average of the ultimate rate of infiltration is always greater than the rate of ultimate infiltration based on the average value of the scale parameter λ.

For the case of the sorptivity, the algebra becomes a little more complex. In this case we have

$$E[S(\lambda)] = \frac{\displaystyle\int_o^\infty p(\lambda)\lambda^{3/2} d\lambda}{(\lambda)^{1/2}} S[E(\lambda)] \tag{24}$$

in which the result is clearly not distribution-free. For the case of twin columns of equal areal extent but contrasting values of λ, as specified by eq. (17), the application of eq. (24) gives

$$E[S(\lambda)] = \frac{(\lambda_1)^{3/2} + (\lambda_2)^{3/2}}{2(\lambda_1/2 + \lambda_2/2)^{3/2}} S[E(\lambda)] \tag{25a}$$

which can be written in terms of $CV(\lambda)$ by use of eq. (17) as

$$E[S(\lambda)] = \frac{[\{\sqrt{2} + CV(\lambda)\}^{3/2} + \{\sqrt{2} - CV(\lambda)\}^{3/2}]}{(2)^{7/4}} S[E(\lambda)] \tag{25b}$$

The values given by eq. (25) for a range of values of $CV(\lambda)$ are shown in the second column of Table 2.

For the gamma distribution as given by eq. (18) the result is given by

$$\frac{\bar{S}(\lambda)}{S(\bar{\lambda})} = \frac{\Gamma(b + 3/2)}{\Gamma(b)} \frac{1}{(b)^{3/2}} \tag{26}$$

Table 2. *Values given by eq. (25) for a range of values of the coefficient of variation CV(λ) for various cases of horizontal hydraulic conductivity K_{hor}.*

| | Value of $E[S(\lambda)]/S[E(\lambda)]$ | | |
| | $K_{hor} = 0$ | | $K_{hor} = \infty$ |
CV	bimodal	gamma	all distributions
0.0	1.000	1.000	1.000
0.2	1.008	1.015	1.020
0.4	1.030	1.058	1.077
0.6	1.068	1.128	1.166
0.8	1.123	1.220	1.281
1.0	1.194	1.329	1.414

The result is seen to depend only on the parameter *b* which is related to the coefficient of variation of the scaling parameter λ through eq. (19d). The results for the gamma distribution model for various values of the coefficient of variation of the scaling parameter λ are given in the third column of Table 2. It can be seen from the second and third columns of this table that the average sorptivity is always less than the sorptivity corresponding to the average scale parameter, but that it is not very sensitive to the distribution used, even for the contrasting case of an extreme U-shaped distribution and a unimodal distribution.

4.2 Limiting case of high horizontal conductivity

The limiting case of infinite horizontal hydraulic conductivity lends itself to relatively simple analysis. The key to this analysis is that under conditions of very high conductivity it will be impossible to maintain a head gradient in the horizontal. This means that for this particular case we have

$$h(x, y, z, t) = h(z, t) \tag{27}$$

which suggests that it may be possible to solve this case in the same way as the one-dimensional case by using the capacity form of the continuity equation as given by eq. (10). It follows from eq. (27) $h(x, y, z, t)$ with respect to time will also be the same in any horizontal plane so that averaging across any arbitrary horizontal plane will give the lump continuity equation at the field scale as

$$\frac{\delta h}{\delta t} E[C(h)] + \frac{\delta}{\delta z} [E\{q(z, t)\}] = 0 \tag{28}$$

Similarly the dynamic equation given by

$$q(x, y, x, t) = -K(h, x, y) \frac{\delta}{\delta z} h(z, t) + K(h, x, y) \tag{29}$$

can be averaged because the gradient with respect to depth of the matric head will be constant at any horizontal surface and we can write

$$E[q(z, t)] = -\frac{\delta}{\delta z} [h(z, t)] E[K(h)] + E[K(h)] \tag{30}$$

For the simple model of eq. (20) on which this partial analysis is based, the capacity $C(h)$ and the conductivity $K(h)$ vary in the horizontal but do not vary with moisture content. Accordingly, the mean rate of infiltration will be given by

$$E[f(t)] = \left\{ \frac{E[K(\lambda)]/E[C(\lambda)]}{\pi t} \right\}^{1/2} + E[K(\lambda)] \tag{31}$$

Since the hydraulic conductivity is proportional to the square of the scaling parameter λ and the capacity C is proportional to λ we have the relationship

$$E[S(\lambda)] = \frac{[E(\lambda^2)/E(\lambda)]^{1/2}}{[E(\lambda)]^{1/2}} S[E(\lambda)] \tag{32a}$$

which can be expressed as

$$E[S(\lambda)] = [1 + \{CV(\lambda)\}^2]^{1/2} S[E\lambda)] \tag{32b}$$

The above expression can be evaluated by equating the coefficient of variation of the assumed distribution to that of the prototype system. The results are shown in column 4 of Table 2.

4.3 Adjustment for atmosphere controlled phases

For the simple model of constant hydraulic diffusivity D and constant hydraulic conductivity K which is taken for both bounds to the analytical solutions, the time for ponding for a precipitation rate P is given by

$$t_p = \frac{\pi}{4} \frac{D}{P^2} \tag{33}$$

For the case of the alternative upper bound in the form of δ-function hydraulic diffusivity and constant hydraulic conductivity, the time to ponding is similar in form and given by

$$t_p = \frac{2}{\pi} \frac{D}{P^2} \tag{34}$$

Since the hydraulic diffusivity is directly proportional to the scaling parameter λ for both these cases, we have

$$\frac{E[t_p(\lambda)]}{t_p[\lambda]} = \frac{E[D(\lambda)]}{D[E(\lambda)]} \tag{35}$$

so that the use of the average value D, based on an assumption for or an estimation of the distribution of the scaling parameter λ, can be inserted in eq. (33) or eq. (34) to obtain the average time to ponding for the case of zero hydraulic conductivity.

For the case of infinite hydraulic conductivity the value of the matric head h will be constant at the surface and hence ponding will occur at the same instant in all parts of the surface plane. This value is given by

$$t_p(\lambda) = \frac{E[K]/E[C]}{E[\lambda]} t_p[E(\lambda)] \qquad (36a)$$

which is equivalent to

$$E_p(\lambda) = [1 + \{CV(\lambda)\}^2] t_p(\lambda) \qquad (36b)$$

A comparison of eqs. (35) and (36) would indicate that for intermediate cases the mean time of ponding would always exceed the time for ponding based on the mean value of the scaling parameter λ.

In applying a time shift to the classical solution for ponded infiltration similar relationships are obtained. The volume infiltrated, according to the classical ponded infiltration case and the simple model of constant D and constant K, is given by

$$F(t) = \left(\frac{4Dt'}{\pi}\right)^{1/2} \qquad (37)$$

where t' is the time elapsed since the onset of infiltration. Since the volume infiltrated for the flux condition is given by

$$F(t) = P\, t_p = \frac{\pi D}{4P} \qquad (38)$$

the combination of eqs. (33), (37) and (38) gives as the value of t'

$$t' = \left(\frac{\pi}{4}\right)^2 t_p \qquad (39)$$

For the case of the δ-function solution a similar exercise leads to the results

$$t' = \left(\frac{2}{\pi}\right) t_p \qquad (40)$$

It will be seen that the ratio of the duration of equivalent ponded infiltration to flux boundary infiltration during ponding is similar for these two widely different models. Again it is clear that allowance for spatial variability is far more important than the differences between models.

5. CONCLUSIONS

Certain conclusions can be drawn from this partial analysis based on the sorptivity-type solution to the problem of infiltration into a semi-infinite soil column. The selection of this special case appears justified since its difference from the other limiting case of an abrupt wetting front is seen to be far less than the usual degree of spatial variability.

The ratio of the spatially averaged value of the ultimate rate of infiltration to the corresponding rate based on the average scaling factor is shown to be greater than unity and to depend only on the coefficient of variation of the scaling and thus to be independent of the type of distribution used. The corresponding ratio of the average value of the sorptivity to the value of sorptivity based on the average scaling parameter is for the lower limit of zero horizontal conductivity insensitive to the type of distribution of the scaling parameter. Comparative results for two widely different distributions and a number of values of the coefficient of variation of the scaling parameter are shown in Table 2.

In regard to the adjustment for pre-ponding conditions in both limiting models, the spatial average values of the time to ponding and of the time of equivalent ponded infiltration for same total volume are both shown to be the same as the corresponding times based on the average value of the scaling parameter.

A good first approximation should be provided by using at a field scale a constant sorptivity based on the expected value of the three-halves power of the scaling parameter and a constant ultimate rate of infiltration based on the expected value of the square of the scaling parameter.

REFERENCES

BEVEN, K.J. and KIRKBY, M.J. (1979). A physically based variable contributing area model of basin hydrology, Hydrol. Sci. Bull., 24(1): 43–69.

BRAESTER, C. (1973). Moisture variation at the soil surface and the advance of the wetting front during infiltration at constant flux, Water Resour. Res., 9:678–694.

CARSLAW, H.S. and JAEGER, J.C. (1946). Conduction of heat in solids. Oxford University Press, Oxford.

CHILDS, E.C. (1936). The transport of water through heavy clay soils. J. Agr. Sci., 26: 114–141, 527–545.

CLOTHIER, B.E., KNIGHT, J.H., and WHITE, I. (1981). Burger's equation: application to field constant-flux infiltration. Soil Sci., 132: 255–261.

DOOGE, J.C.I. (1986). Scale problems in hydrology. Fifth Chester C. Kisiel Memorial Lecture, Department of Hydrology and Water Resources. Univ. Arizona, Tucson.

DOOGE, J.C.I. and WANG, Q.S. (1993). Comment on 'An investigation of the relationship between ponded and constant flux rainfall infiltration' by A. Poulavassilis et al., Water Resources Research, 39(4): 1335–1337.

FUJITA, H. (1952). The exact pattern of concentration dependent diffusion in a semi-infinite medium, Part II, Textile Res. J., 22: 823–827.

GREEN, W.H. and AMPT, G.A. (1911). Studies on soil physics. I. Flow of air and water through soils. J. Agri. Sci., 4: 1–24.

HOPMANS, J.W. (1987). A comparison of various methods to scale soil hydraulic properties, J. Hydrol., 93: 241–256.

KLINE, S.J. (1965). Similitude and approximating theory. McGraw Hill, New York.

KNIGHT, J.H. and PHILIP, J.R. (1974). Exact solutions in non-linear diffusion. J. Eng. Maths, 8: 219–227.

KUHNEL, V., DOOGE, J.C.I., O'KANE, J.P.J. and ROMANOW-ICZ, R.J. (1991). Partial analysis applied to scale problems in surface moisture fluxes. Surveys in Geophysics, 12: 221–241.

MEIN, R.G. and LARSON, C.L. (1973). Modelling infiltration during a steady rain. Water Resour. Res. 9: 384–394.

MILLER, E.E. and MILLER, R.D. (1955). Theory of capillary flow I. Practical implications, Soil. Sci. Soc. Amer. Proc., 19: 267–271.

MILLER, E.E. and MILLER, R.D. (1956). Physical theory for capillary flow phenomena. J. Appl. Phys., 27: 324–332.

MILLY, P.C.D. and EAGLESON, P.S. (1982). Infiltration and evaporation at inhomogeneous land surfaces, Report No. 278, Department of Civil Engineering, MIT, Cambridge, Mass.

PHILIP, J.R. (1954). An infiltration equation with physical significance, Soil Sci., 77: 1153–1157.

PHILIP, J.R. (1957a). The theory of infiltration: 1. The infiltration equation and its solution, Soil Sci., 83: 345–357.

PHILIP, J.R. (1957b). The theory of infiltration: 4. Sorptivity and algebraic infiltration equations, Soil Sci., 84: 257–264.

PHILIP, J.R. (1958). The theory of infiltration, 7, Soil Sci., 85: 333–337.

PHILIP, J.R. (1966). A linearization technique for the study of infiltration. In: R.E. Rijtema and H. Wassink (Editors) Proc. IASH/UNESCO Symp. on Water in the Unsaturated Zone. UNESCO, Paris, Vol 1: 471–478.

PHILIP, J.R. (1969). Theory of infiltration, in V.T. Chow (editor) Advances in hydrosciences, 5: 216–296, Academic Press, New York.

PHILIP, J.R. (1974). Recent progress in the solution of nonlinear diffusion equation. Soil Sci., 117: 257–264.

POLYA, G. (1957). How to solve it, Doubleday and Co., Garden City, New York.

PRANDTL, L. (1904). Uber Flüssigkeitsbewegung bei sehr kleiner Reibung, Proc. 3rd Intern. Math. Congr., Heidelberg.

ROGERS, C., STALLYBRASS, M.P., and CLEMENTS, D.L. (1983). On two-phase infiltration under gravity and with boundary infiltration: An application of a Backlund transformation. Nonlinear Anal. Theory Math. Appl., 7: 785–799.

ROMANOWICZ, R.J., DOOGE, J.C.I. and O'KANE, J.P.J. (1990). Partial analysis at field and biome scale. EV4C-0076-IRL Project Report. Centre for Water Resources Research, University College Dublin.

SANDER, G.C., PARLANGE, J. -Y., KÜHNEL, V., HOGARTH, W.L., LOCKINGTON, D. and O'KANE, J.P.J. (1988). Exact nonlinear solution for constant flux infiltration. J. of Hydrol., 97: 341–346.

VAUCLIN, M. (1984). Infiltration in unsaturated soils in J. Bear and M.Y. Corapcioglu (editors), Fundamentals of Transport Phenomena in Porous Media, Martinus Nijhoff, Dordrecht, 257–313.

WANG, Q.J. and DOOGE, J.C.I. (1993). Limiting cases of land surface fluxes. XVI General Assembly of European Geophysical Society, Wiesbaden, April, 1991. J. of Hydrology, 155(1994): 429–440.

WOOD, E.F., SIVALAPON, M., BEVEN, K.J. and BAMD, L. (1988). Effect of spatial variability and scale with implications for hydrologic modelling. J. of Hydrology, 102: 29–47.

Remote sensing – inverse modelling approach to determine large scale effective soil hydraulic properties in soil–vegetation–atmosphere systems

R. A. FEDDES

Department of Water Resources, Wageningen Agricultural University,

Nieuwe Kanaal 11, 6709 PA Wageningen, The Netherlands

ABSTRACT This paper gives a general overview of how parameter optimization using inverse modelling can be used to infer scale-dependent effective soil hydraulic functions. In this approach, it is assumed that macroscopic flow behaviour as represented by Richards' equation is valid at all spatial scales. Thus, solution of the continuity equation with appropriate boundary conditions is independent of the size of the physical system. In this manner, for example, watershed effective soil hydraulic properties can be estimated. Since the optimization requires areal-averaged boundary conditions and measured areal-averaged flow attributes, such as water content, it is shown how remote sensing can be applied to infer these area-averaged hydrological variables.

In this way effective soil hydraulic properties are obtained from inverse modelling using the dynamical one-dimensional soil–water–vegetation model SWATRE. Areal evaporation can be obtained from reflection and thermal infrared remote sensing, while areal soil water content is estimated from microwave remote sensing techniques. The combined remote sensing and inverse modelling approach is illustrated with data taken from the HAPEX-EFEDA experiment in Spain and from the catchment Hupselse Beek in the Netherlands. It is concluded that small-scale soil physics may adequately describe mesoscale behaviour.

1. THE PROBLEM OF SCALES

The atmospherical, hydrological and terrestrial components of the earth system operate on different time and space scales. Resolving these scaling incongruities as well as understanding and modelling the complex interaction of land surface processes at the different scales represent a major challenge for hydrologists, ecologists and meteorological scientists alike.

Studies of the soil water balance are usually done at a spatial scale of 1 to 100 m, using a time scale of days to months. Micrometeorological studies are carried out over the same spatial scale but with a much shorter time resolution. Mesometeorological studies are carried out over kilometre-size regions, whereas regional hydrological problems are posed in this space scale or larger, but with a time resolution of months to years. Physical and biological processes in the atmospheric boundary layer and at the land-atmosphere interface interact non-linearly and at different scales.

The link between Global Change Models (GCMs) must serve in two ways (Shamir, 1992). First, terrestrial processes must be represented adequately in GCMs. Probably the weakest part of GCMs today is the link with the lower boundary at the soil surface. To provide this link it is necessary to bridge across the different scales used in hydrology and GCMs. The hydrological models must also be simple enough, so as not to create an unacceptable computational burden. Second, the combined model should be able to reflect variations in soils, topography and vegetative cover. The connection between hydrology and meteorology occurs through the planetary boundary layer, so that the 'missing link' between GCMs and hydrological models manifests itself at the mesoscale.

To model the effect of changes in land surface heterogeneity on the terrestrial water balance, one often applies semi-

distributed models. In this type of model the spatial variabilities of input, parameters and output are accounted for in
either a deterministic or statistical way. Alternatively, one
may consider lumped (macroscale) type of models, which
view the input, parameters and output to be homogeneous in
space, i.e. as a one-dimensional model. As scale increases the
difference in hydrological response usually tends to decrease.
Wood (this volume) shows that one may refer to a certain
threshold scale ('elementary representative area') at which
the variance is at a minimum: i.e., at a scale 2 to 3 times the
correlation length of the process, or about 1 to 2 km^2 for a
natural catchment.

One-dimensional lumped models such as the vertical Soil–
Vegetation–Atmosphere–Transfer schemes (SVATs) may,
however, fail when lateral transport of water by overland or
subsurface flow occurs. Becker (this volume) and Hatton et
al. (this volume) show that for catchments with complex
sloping terrains and groundwater tables, a vertical domain
SVAT has to be coupled with either a process or a statistically
based scheme that incorporates lateral transfer.

Kite et al. (this volume) represent catchment heterogeneity
by identifying a limited number of important land use types
using remote sensing techniques and aggregation. Each land
use group is then represented in a streamflow routing model
by three non-linear reservoirs for snow, surface water and
groundwater, respectively. These reservoirs are intercon-
nected to correspond to the physical layout of the watershed.
In this hierarchical approach of Kite et al. a top level water
balance model at the GCM grid scale can be applied that
interacts with the input/output of a GCM. The time step of
the flow routing model is on an hourly basis, of the watershed
model on a daily basis and that of the top level on a monthly
basis.

A problem in most hydrological models is that they do not
consider the lateral transfer of water and energy in the air
over heterogenous land surfaces, because of the poor understanding of their interactions with the planetary boundary
layer (e.g. Raupach, 1991)

This paper focuses on the estimation of regional effective
soil hydraulic parameters by inverse one-dimensional modelling. Generally speaking, one may state that a physically-
based model for vertical soil water movement is required that
reacts to changes in boundary conditions, irrespective of the
spatial scale considered. Soils, because of their non-linear
flow properties transform sudden changes in boundary conditions into gradual changes in water movement. These
macroscopic flow properties consist of the relationships
between pressure head, soil water content and hydraulic
conductivity. If we assume that existing small-scale soil
hydraulic property formulations can adequately describe the
large-scale hydraulic properties, then traditional small-scale

soil physics may be employed to estimate areal vertical water
movement at larger scales, i.e., by application of the one-
dimensional Richards' flow equation.

Based on this reasoning, a new approach is explored to
calculate efficiently average vertical soil water movement for
large areas. If successful, the approach will eliminate the need
to investigate the hydrological behaviour on separate plots
within the larger area of investigation. The basic idea is to
derive, by inverse modelling, effective soil hydraulic properties for an area as a whole, using *areal flow data* (e.g., areal
evaporation, areal drainage, areal water storage changes). In
principle, the inverse method allows the area to be modelled
for a specific boundary condition by one single simulation of
a one-dimensional model. In this way, spatial variability is
averaged before instead after model simulations.

The method extends the use of inverse modelling from
plots to large areas. The soil hydraulic properties that are
fitted to areal water movement data in combination with the
Richards' equation are thus assumed to describe effectively
the hydrological behaviour of that area. In this inverse
modelling approach one needs, besides the initial and lower
boundary conditions, information about the upper boundary condition, i.e., areal evapotranspiration, and also about
the areal soil moisture content. These properties can nowa-
days be inferred from airborne and satellite remote sensing.
This combined inverse modelling-remote sensing approach is
illustrated with data taken from the slightly sloping catchment Hupselse Beek, which shows a great variety in soil
hydraulic properties, and with data from the HAPEX-
EFEDA experiment in Spain.

2. MODELLING THE SOIL–WATER–
VEGETATION–ATMOSPHERE SYSTEM —

The one-dimensional soil water balance equation can be
expressed by the differential form of the continuity equation
according to:

$$\frac{\partial \theta}{\partial t} = -\frac{\partial q}{\partial z} - S \qquad (1)$$

where q is the volumetric (Darcy) flux (cm d^{-1}) and S
represents the volume of water taken up by the roots per unit
bulk volume of soil per unit time (d^{-1}). Eq. (1) can be written
in terms of soil water pressure head h (cm). As $\partial \theta/
\partial t = C(h)(\partial h/\partial t)$, with C being the differential soil water
capacity $d\theta/dh$ (cm^{-1}), and $q = -K(h)[\partial h/\partial z + 1]$, with K
being the hydraulic conductivity (cm d^{-1}) and the vertical
coordinate z (cm) taken positive upwards, eq. (1) reads as:

$$\frac{\partial h}{\partial t} = \frac{1}{C(h)} \frac{\partial}{\partial z} \left[K(h) \left(\frac{\partial h}{\partial z} + 1 \right) \right] - \frac{S}{C(h)} \qquad (2)$$

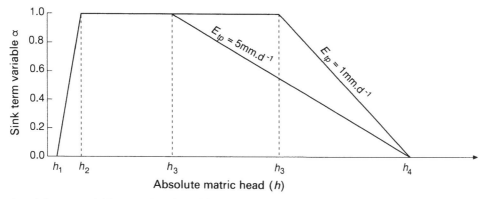

Fig. 1 Dimensionless sink term variable α as a function of the absolute value of the soil water matrix head $|h|$ (after Feddes et al., 1978). Water uptake below $|h_1|$ (oxygen deficiency) and above $|h_4|$ (wilting point) is set at zero. Between $|h_2|$ and $|h_3|$ (reduction point) water uptake is maximal. The value of $|h_3|$ varies with the potential transpiration rate E_{tp}.

Eq. (2) is known as the Richards equation. Feddes et al. (1978) described S semi-empirically by

$$S(h) = \alpha(h) S_{max} \qquad (3)$$

where $\alpha(h)$ is a dimensionless prescribed function of the pressure head and S_{max} is the maximal possible water extraction by roots. In the interest of practicality a homogeneous root distribution over the soil profile is assumed and S_{max} is defined as:

$$S_{max} = \frac{E_{tp}}{|z_r|} \qquad (4)$$

where E_{tp} is the potential transpiration rate and $|z_r|$ is the rooting zone depth. For alternative root water uptake expressions see Feddes et al. (1988).

So far root water uptake under optimal soil water conditions, S_{max} has been considered. Under non-optimal conditions, i.e., either too dry or too wet, S_{max} is reduced by means of the $\alpha(h)$ function. The shape of this function is shown in Fig. 1.

For a detailed description of the initial, upper and lower boundary conditions of eq. (2), the reader is referred to Feddes et al. (1988). Potential transpiration and potential soil evaporation are used as forcing functions and depend on atmospherical conditions and the type (canopy resistance) and amount (leaf area index) of vegetation. The numerical model SWATRE (Belmans et al. 1983) has been developed to solve eq. (2) for different soil depths using variable time steps. The output of the model includes the soil water flux at each depth, actual transpiration (computed as the integral of the sink term over the rooting depth), soil evaporation/infiltration and the flux at the bottom of the soil profile.

A model like SWATRE may be used to calibrate or to improve the hydrological parameterizations applied in meso- and large-scale atmospheric models (e.g. Stricker et al., 1993; Kim, 1993).

3. ESTIMATION OF AREAL EVAPOTRANSPIRATION FROM REFLECTION AND THERMAL INFRARED REMOTE SENSING

Since the 1970s, a great deal of research has been devoted to developing methods to obtain geophysical variables at the land surface by means of space observations. The potential of space measurements to study land surface processes has often been demonstrated (Schmugge and Becker, 1991; Choudhury, 1991). Surface soil moisture, surface reflectance and surface temperature can be estimated with satellite images. Shortcomings still exist, however. A complex variable, such as evapotranspiration flux, cannot be measured directly from space, but can be estimated from simplified models that describe the transfer of energy, fluxes and momentum at the land surface using remotely detectable quantities.

Inversion procedures commonly used for the translation of satellite images assume independence between variables such as surface temperature, reflectance, vegetation index and aerodynamic roughness length. Although this might be correct locally and when considering small changes, it is not true when large changes in space and time are considered. Cross-correlation analysis of estimates of, for example, surface reflectance and temperature has been reported in the literature (Menenti et al., 1989). It appears that such correlation is a dominant characteristic of the combined complex land surface processes which partition the available energy at the land surface. In particular, Bastiaanssen (1991) has shown that empirical regression coefficients obtained by relating surface radiation temperature to surface reflectance can be applied to obtain values of effective aerodynamic resistances.

To obtain estimations of actual evapotranspiration, additional data or simplifying assumptions are thus needed. Formally, the problem can be formulated as follows.

The energy balance for a homogeneous surface can be written as:

$$Q^* = H + G_0 + \lambda E \quad (\mathrm{W\ m}^{-2}) \tag{5}$$

where Q^* is the net radiation flux, H the sensible heat flux, G_0 the soil heat flux and λE the latent heat flux.

Eq. (5) can be rewritten as a function relating λE to the other variables:

$$\lambda E = (1 - r_0) K\downarrow + \varepsilon' \sigma T_a^4 - \varepsilon_0 \sigma T_0^4 - \frac{\rho_a c_p}{r_{ah}} (T_0 - T_a)$$
$$- \frac{\rho_s c_s}{r_{sh}} (T_0 - T_s) \tag{6}$$

where $r_0 (-)$ is the surface reflectance (remotely sensed), $K\downarrow$ $(\mathrm{W\ m}^{-2})$ is the global solar radiation, $\varepsilon' \sigma T_a^4$ $(\mathrm{W\ m}^{-2})$ is the long-wave sky emittance with $\varepsilon' (-)$ being the apparent emissivity of the atmosphere, σ $(\mathrm{W\ m}^{-2}\ \mathrm{K}^{-4})$ is the Stefan-Boltzmann constant, T_a (K) is the screen-height air temperature, $\varepsilon_0 \sigma T_0^4$ $(\mathrm{W\ m}^{-2})$ is the long-wave surface emittance with T_0 (K) the surface temperature (remotely sensed), ρ_a (kg m^{-3}) is the air density, ε_0 is the landsurface emissivity, c_p (J $\mathrm{kg}^{-1}\mathrm{K}^{-1}$) is the specific heat of air at constant pressure, r_{ah} (s m^{-1}) is the mean turbulent resistance, ρ_s (kg m^{-3}) is the soil density, c_s (J $\mathrm{kg}^{-1}\ \mathrm{K}^{-1}$) is the soil specific heat, r_{sh} (s m^{-1}) is the soil resistance to heat transfer and T_s (K) is the soil temperature. To obtain eq. (6), it was necessary to invert the heat transport equations of sensible and soil heat flux. Note that with eq. (6) instantaneous evapotranspiration values are obtained.

In order to assess areal patterns of the variables of eq. (6), an Surface Energy Balance ALgorithm (SEBAL) has been developed by Bastiaanssen (1993). The algorithm is used in connection with empirical relationships for those variables that are not directly observable by satellites. By remote sensing Bastiaanssen (1993) determined the Normalized Difference Vegetation Indexes (NDVI), $K\downarrow$, r_0 and T_0 for an area around Barrax in Castilla la Mancha in Spain. From combination of these data with field measurements he derived empirical relationships between the remotely sensed data and the parameters T_a, z_0, ε_0 and G_0/Q^*. Then pixelwise values of NDVI, $K\downarrow$, r_0, T_0, G_0/Q^*, ε', ε_0, aerodynamic roughness length, z_0, friction velocity $\mu*$ and aerodynamic resistance r_a were mapped. After separate computation of O^*, G_0 and H, values for λE were obtained from the residual of the surface energy balance. For full details of the procedure see Bastiaanssen and Roebeling (1993), Bastiaanssen et al. (1993), and Feddes et al. (1993b).

As an example of this approach in Fig. 2 the frequency distributions of the terms of the energy balance on 29 June 1991 are shown for an area 15.9 km × 14.4 km around

Barrax. Some 35% of the area consists of irrigated land. The wide diversity of the hydrological (wet and dry) conditions in the area is clearly reflected in the four frequency distributions.

The wide range in net radiation values, Q^*, reveals a wide diversity of the land hydrological conditions. The low soil heat fluxes, G_0, represent areas with a high leaf area index while the higher values represent the bare rain-fed parcels. The sensible heat fluxes, H, vary considerably, with some high peaks observed above dry tall canopies. The latent heat fluxes, λE, are log-normally distributed, with high values originating from the irrigated fields. These types of distribution indicate the relative homogeneity of the investigated area.

4. ESTIMATION OF AREAL SOIL MOISTURE FROM MICROWAVE REMOTE SENSING

Active microwave remote sensing of soil moisture relies on the large contrast between the dielectric constants of water and dry soil. Passive microwave radiometers measure the thermal emission from the surface. Applying L-band radiometry, the latter method has up to now mainly been used for soil moisture estimation of the upper 10 cm soil layer. Soil moisture estimates over larger depths require a-priori knowledge of soil type and hydrological situation.

Active microwave techniques, and especially multi-frequency radar, provide areal information on soil moisture content with depth and its variability. The penetration depth of the electromagnetic wave depends on the frequency band and the actual soil moisture content. Scattering of electromagnetic waves by rough soil and vegetation surfaces can be described by relatively simple models. Examples are the small perturbation model for 'smooth' surfaces, the physical optics model for 'intermediate rough' and the geometrical optics model for 'very rough' surfaces (e.g. Ulaby and Elachi, 1990). When using polarimetric data the HH/VV polarization ratio can be applied, without the need for exact knowledge of surface roughness (Shi et al. 1991). To relate the soil dielectric constant to the soil moisture content various dielectric mixing models are available ranging from empirical to more physically-based approaches (Wang and Schmugge, 1980; Dobson et al., 1985).

Bastiaanssen et al. (1994) used the multi-frequency Synthetic Aperture Radar (SAR) with a C-band (5 cm), an L-band (20 cm) and a P-band (68 cm) to derive the soil moisture content under different types of land use on 19 June 1991 in the Barrax area of Spain. These moisture data were compared with gravimetrically determined soil moisture contents (Table 1).

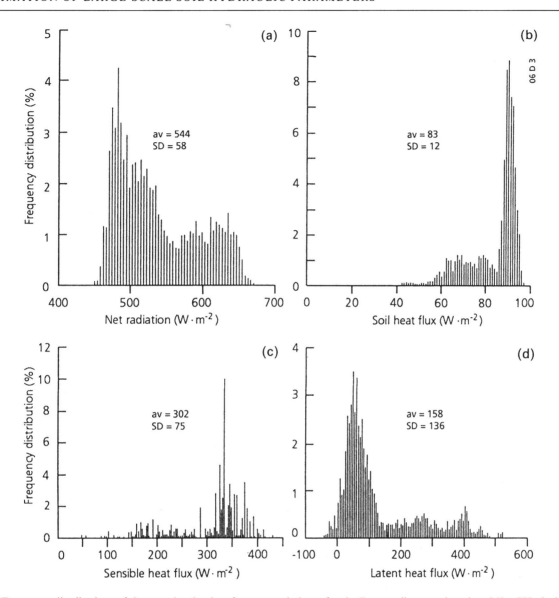

Fig. 2 Frequency distributions of the complete land surface energy balance for the Barrax pilot area, based on 862 × 777 pixels with a size of 18.5 m × 18.5 m each (after Bastiaanssen et al. 1993). (a) Net radiation (Q^*); (b) soil heat flux density (G_o); (c) sensible heat flux density (H); (d) latent heat flux density (λE).

Although the microwave backscatter bare soil model of Oh et al. (1992) used in this case has a rather limited applicability for vegetated plots, the results look promising. For the bare soil, maize and barley plots, the water content data of the L-band (penetration depth 40 cm) compare favourably with those measured gravimetrically at depths of 5, 10 and 30 cm. The higher moisture contents in the P-band are related to the larger penetration depth as compared to the other two bands. This depth varies from about 1.65 m for a completely dry soil to about 0.20–0.50 m for wet to moist soils.

A multi-frequency instead of a single frequency, multi-polarized approach and a better correction algorithm for vegetation influence may further improve the results as well as increase the applicability of soil moisture estimators. It

thus may be concluded that a combination of multi-temporal microwave observations with field data on surface soil moisture has great potential for indirect estimation of areal soil hydraulic properties.

5. ESTIMATION OF LARGE-SCALE EFFECTIVE SOIL HYDRAULIC PARAMETERS BY INVERSE MODELLING OF AREAL EVAPOTRANSPIRATION AND AREAL SOIL MOISTURE

5.1 General

Given areal estimates of both areal evaporation fluxes and soil moisture at different depths, it becomes attractive to

Table 1. *Volumetric soil moisture contents as determined by JPL-SAR in the L- and P-bands on 19 June 1991 on various field plots in Barrax, Spain as compared with gravimetrically measured data (after Bastiaanssen et al. 1994).*

Depth (cm)	Soil moisture content $\theta(-)$		
	Bare soil	Maize (irrig.)	Barley (irrig.)
2	—	0.130	—
5	0.034	0.186	0.054
10	0.071	0.261	0.097
20	—	—	0.122
25	—	—	0.140
30	0.160	—	—
L-band	0.11	0.13	0.10
P-band	0.18	0.17	—

estimate the soil's areal average hydraulic properties over large areas. However, the large spatial variability of the hydraulic properties of soils makes the definition and estimation of 'average' soil hydraulic properties extremely difficult. Even a texturally homogeneous soil type may exhibit large spatial variability. It is thus inappropriate to use point estimates of soil hydraulic properties or their simple mean value as areal representative values. Aggregation from local to large scale (regionalization) can generally be performed by scaling, stochastic analysis or inverse modelling.

In the scaling approach the hydraulic functions such as the soil moisture characteristic $\theta(h)$ and the hydraulic conductivity relationship $K(\theta)$ of eq. (2) can be related to reference curves by single valued scaling factors. Knowing the reference curve and the statistical distribution of the scaling factor the Monte-Carlo technique can be used to generate soil hydraulic properties (e.g., Hopmans and Stricker, 1989). To arrive at areal hydrological values, the results of many simulations of one-dimensional soil water flow, each representing part of the area, are averaged.

Many studies have been based on a perturbation expansion of the Richards equation often necessitating simplifying assumptions. These analytical solutions imply, for example, that the statistical variation of the soil hydraulic properties and boundary conditions fluctuates in space around constant means with their fluctuations having constant higher statistical moments (variance, covariance, skewness, third order correlation, etc.) in space. Chen et al. (1993) developed a second order closure form of the areally averaged Richards equation and a complete closure form of the areally averaged Green-Ampt model. They also incorporated the saturated hydraulic conductivity as a stochastic field with specified mean and covariance structure to account for large spatial variabilities. The study of Chen et al. (1993) has been confined to infiltration in a homogeneous soil profile

with gravity flow at the bottom. Hence, it would be interesting to extend their approach to heterogenous soil profiles with spatially varying rain falls, evaporation and transpiration upper boundaries, and a (shallow) groundwater table at the lower boundary.

The inverse modelling approach is receiving much attention. It treats the actual heterogeneous soil as an equivalent homogeneous system with a set of effective properties. Discussions on the existence and physical meaning of large-scale effective soil physical parameters and on their applicability to large-scale hydrology are still on going (e.g. Beven, 1989). Yet, several studies have shown that effective soil physical parameters can be derived. When applied through a numerical model to a homogenized soil system, the response in terms of, for example, evaporation and percolation is comparable to that of the actual heterogeneous system (Mantoglou and Gelhar, 1987; Mishra et al., 1990). A conditional probability-based approach for estimating grid scale effective parameters from a limited number of available measurements has been presented recently by Beven (1993).

The approach of Kool and Parker (1987) is based on solution of the inverse problem using a parameter estimation technique in combination with an optimization procedure. In this approach, the flow problem may be formulated for any set of initial and boundary conditions and be solved with any analytical or numerical model. Input to such a model are measured water contents and/or pressure head distributions and/or outgoing and incoming water fluxes. While assuming a particular form for the soil hydraulic functions, the parameters in those functions are estimated by using an optimization procedure that minimizes an objective function, containing the predicted and measured flow variables.

Only indirect inverse methods which are based on the minimization of the output error optimize the predictive model performance. In other words, these *indirect inverse methods are the only ones that actually solve the flow equations for head values and fluxes of a system in relation to system-dependent parameters.* Going from the scale of a representative elementary volume (e.g. Dooge, this volume) to larger scales, it is possible to assign to the parameters of such larger systems 'effective qualities'. These 'effective qualities' will be scale-dependent, i.e., one would obtain different values and ranges for different 'lumped' scales to which the indirect inverse estimation method is applied. Of course one might question the validity of the Richards equation on such lumped scales. However, none of the existing alternatives gives a better conceptual basis.

5.2 Inverse modelling approach of a catchment: simulation experiment

In principle the inverse method allows the hydrological modelling of any size area by a single simulation of the one-

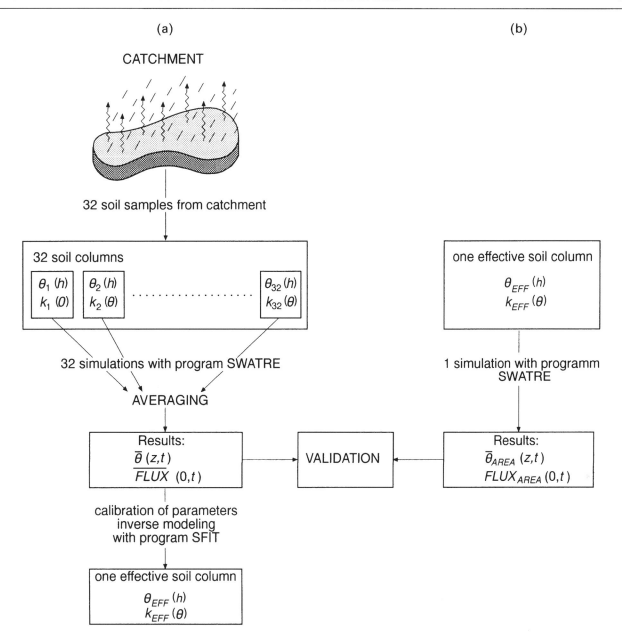

Fig. 3 (a) Scheme of the numerical simulation and the inverse modelling procedure followed for the fitting case. (b) Scheme of the validation procedure followed for the fitting, wet and dry cases (after Feddes et al., 1993a).

dimensional model in eq. (2), as the spatially averaged input and output are used in the model.

In the study reported by Feddes et al. (1993a), the feasibility of deriving field-effective parameters by large-scale inverse modelling was investigated. Numerical evaporation-infiltration experiments by the SWATRE model were simulated on 32 hypothetical bare-soil columns having different hydraulic properties. The latter were taken from 32 samples of the top layer of an existing fine sandy soil in the Hupselse Beek catchment of 6.5 km^2 (Fig. 3(a)).

The outputs of the 32 simulations were averaged to obtain daily areal soil water content profiles $\bar{\theta}(z,t)$ and areal evaporation and infiltration fluxes $\overline{FLUX}(0,t)$. Subsequently the effective soil hydraulic parameters were estimated using the

inverse modelling approach (Kool and Parker, 1987) employing spatially averaged data to construct one single 'effective' soil column, representing the catchment (Fig. 3(b)). The following hydrological data sets were used for the inverse problem:

(1) water content profiles on selected days;
(2) cumulative evaporation and infiltration data;
(3) cumulative evaporation and infiltration data supplemented with surface soil water content at 0.5 cm depth.

Alternatively effective hydraulic parameters, denoted reference hydraulic parameters, were obtained by scaling of the 32 soil samples using a similar media technique (Hopmans and Stricker, 1989).

Table 2. *Effective soil hydraulic parameters resulting from some optimization and scaling procedures (after Feddes et al., 1993a).*

Parameter set[a]	θ_r (−)	θ_s (−)	α (cm^{-1})	n (−)	K_s (cm d^{-1})	ℓ (−)
1	0	0.419	0.0286	1.348	124.6	0.970
3	0.0183	0.398	0.00788	1.386	33.7	0.500
7	0	0.4063	0.0274	1.4072	125.6	0.8799

Notes:

[a] The number of the parameter set refers to the method by which the parameter values were obtained.

    1: Fit on soil moisture profiles, all parameters optimized.

    3: Fit on infiltration and evaporation, all parameters being optimized.

    7: Reference curves obtained from scaling using soil hydraulic data from a wider range of water contents than applied by Hopmans and Stricker (1989).

The soil hydraulic conductivity curves were parameterized by fitting the Mualem-van Genuchten functions (van Genuchten, 1980) to the measured data:

$$S_e = [1 + (\alpha|h|^n]^{-m} \tag{7}$$

$$K = K_s S_e^l \left[1 - (1 - S_e^{1/m})^m\right]^2 \tag{8}$$

where $S_e = (\theta - \theta_r)/(\theta_s - \theta_r)$, $m = 1 - 1/n$, K_s is the hydraulic conductivity at saturation (LT^{-1}), and K is the unsaturated hydraulic conductivity. The subscripts r and s denote the residual and saturated values of the volumetric water content $\theta(-)$, while $n(-)$, $\alpha(L^{-1})$, and $\ell(-)$ are shape parameters.

A selection of the results of the inverse modelling approach of the bare soil columns is shown in Table 2.

5.3 Inverse modelling approach: validation

In the validation runs the parameter sets 1, 3 and 7 of Table 2 were tested on two problems representing wet and dry conditions. The ranges of pressure heads and water contents as covered by the test problems were partly beyond the range over which the parameters were fitted. The initial condition for the runs employing the different parameter sets in both cases was the average of the initial soil moisture profiles of the 32 artificial soil columns.

It was found (for full details, see Feddes et al., 1993a) that optimization on areal fluxes (parameter set 3) does not give satisfactory results, while optimization on average soil moisture profiles (parameter set 1) does. The reference parameters obtained from scaling (parameter set 7) performed equally as well as parameter set 1.

The derived effective hydraulic parameter sets were vali-

dated for artificial dry and wet conditions. In addition the parameter sets 1, 3 and 7 of Table 2 were judged by simulating the water balance of a grass crop under natural meteorological conditions. If a parameter set is valid under extreme, artificial conditions, it probably will also be valid under less extreme, but natural conditions. On the other hand, parameter sets that deviate under extreme conditions, might show acceptable results when simulating moderate situations.

Growing a grass crop has two effects on the water regime. It causes a higher soil water loss by transpiration as compared to the bare soil. At the same time the $\theta(z, t)$ profile will probably be more homogenized, because of water extraction by roots from relatively wet soil depths.

The simulations were performed with the SWATRE model for the growing season of the fairly dry year 1982 and the extremely dry year 1976, as these years show the largest variation in pressure head profiles with depth in time.

Simulations were again performed for the same 32 homogeneous soil columns of the Hupselse Beek catchment area. On the first day of each month the mean and standard deviation of cumulative actual transpiration, cumulative flow through the bottom boundary, and volumetric water contents at 1.5 and 32.5 cm depth were calculated. These quantities were selected because of their relevance for crop growth, recharge to the groundwater table and runoff generation. The results were compared with simulations using the effective soil hydraulic functions of Table 2.

Table 3 gives a summary of the SWATRE output. The years 1976 and 1982 show similar trends for all the parameter sets. The order of best performance is given in Table 4. Actual transpiration and flow to the groundwater are accurately simulated by both parameter set 1 and parameter set 7. The water contents in the top soil are better predicted by parameter set 1. Parameter set 3 shows less extreme water content values than the mean. This was expected as the optimization on evaporation data results in soil hydraulic functions with less steep gradients. Parameter set 7 underestimates the water content in the top soil, although the fluxes at both boundaries are correctly simulated. Hence, the final conclusion is that optimization on soil moisture profiles yields the parameter set which performs best.

CONCLUSION

We might thus conclude that small-scale soil physics may adequately describe large-scale hydrological phenomena. The effective hydraulic parameters describing these phenomena may be derived by inverse modelling on the basis of areal evapotranspiration and surface soil moisture data obtained by remote sensing. As more information on soil moisture at

Table 3. *Summary of the SWATRE simulations of a grass crop over the period 1 April to 1 October of the extremely dry year 1976 and the fairly dry year 1982, applying the effective soil hydraulic functions of Table 2 (after Feddes et al. 1993a).*

	Cumulative actual transpiration (cm)									
	1976					1982				
Daynr	Mean	⟨Std⟩	Set 1	Set 3	Set 7	Mean	⟨Std⟩	Set 1	Set 3	Set 7
120	4.13	0.141	4.16	4.08	4.13	3.81	0.025	3.82	3.77	3.82
150	8.76	0.771	8.82	9.20	8.61	9.48	0.207	9.54	9.25	9.51
180	13.82	1.213	13.91	14.34	13.60	13.81	0.813	13.90	14.07	13.66
210	16.19	1.283	16.26	17.22	15.94	18.89	1.343	18.95	18.88	18.58
240	19.23	1.495	19.34	19.99	18.95	21.13	1.336	21.22	21.64	20.86
270	21.48	1.334	21.63	22.28	21.31	24.65	1.333	14.81	24.82	24.45

	Cumulative flow to the groundwater (cm)									
	1976					1982				
Daynr	Mean	⟨Std⟩	Set 1	Set 3	Set 7	Mean	⟨Std⟩	Set 1	Set 3	Set 7
120	0.85	0.059	0.84	0.75	0.85	0.85	0.054	0.84	0.78	0.85
150	1.24	0.186	1.19	1.05	1.19	1.35	0.145	1.30	1.21	1.30
180	1.44	0.275	1.34	1.05	1.34	1.64	0.291	1.55	1.38	1.55
210	1.56	0.338	1.43	1.09	1.43	1.81	0.291	1.68	1.50	1.68
240	1.64	0.386	1.49	1.12	1.49	1.92	0.334	1.77	1.56	1.77
270	1.70	0.426	1.53	1.14	1.53	2.00	0.370	1.84	1.60	1.83

	Volumetric water content θ at 1.5 cm soil depth									
	1976					1982				
Daynr	Mean	⟨Std⟩	Set 1	Set 3	Set 7	Mean	⟨Std⟩	Set 1	Set 3	Set 7
120	0.077	0.058	0.075	0.123	0.053	0.186	0.072	0.187	0.194	0.159
150	0.221	0.073	0.218	0.201	0.192	0.108	0.067	0.106	0.128	0.080
180	0.064	0.051	0.067	0.068	0.048	0.262	0.075	0.259	0.241	0.231
210	0.165	0.068	0.164	0.145	0.140	0.059	0.048	0.066	0.072	0.048
240	0.064	0.044	0.067	0.080	0.053	0.206	0.071	0.205	0.186	0.179
270	0.144	0.065	0.145	0.131	0.121	0.172	0.067	0.172	0.155	0.148

	Volumetric water content θ at 32.5 cm soil depth									
	1976					1982				
Daynr	Mean	⟨Std⟩	Set 1	Set 3	Set 7	Mean	⟨Std⟩	Set 1	Set 3	Set 7
120	0.166	0.072	0.168	0.168	0.141	0.177	0.073	0.179	0.182	0.151
150	0.138	0.063	0.137	0.158	0.113	0.155	0.070	0.156	0.163	0.129
180	0.127	0.060	0.127	0.121	0.105	0.155	0.056	0.156	0.188	0.137
210	0.122	0.058	0.120	0.126	0.099	0.128	0.060	0.131	0.127	0.108
240	0.115	0.058	0.115	0.105	0.094	0.130	0.054	0.124	0.152	0.104
270	0.112	0.057	0.111	0.117	0.090	0.118	0.058	0.120	0.121	0.098

Notes:

Mean = arithmetic mean of 32 locations in the Hupselse Beek catchment

⟨Std⟩ = standard deviation of the results for the 32 locations

Set 1 = optimized soil hydraulic functions using $\theta(z, t)$ data; parameter set 1 of Table 2.

Set 3 = optimized soil hydraulic functions using evaporation data; parameter set 3 of Table 2.

Set 7 = reference soil hydraulic functions as derived by scaling; parameter set 7 of Table 2.

Table 4. *Accuracy of the results of the SWATRE simulations (see Table 3) with the parameter sets 1, 3 and 7 of Table 2 (after Feddes et al. 1993a).*

Actual transpiration	Set 1 = Set 7 > Set 3
Bottom boundary flow	Set 1 = Set 7 > Set 3
Soil water content θ ($z = -1.5$ cm)	Set 1 > Set 3 > Set 7
Soil water content θ ($z = -32.5$ cm)	Set 1 > Set 3 > Set 7

larger depths becomes available, e.g. by microwave techniques, optimization results may be improved.

ACKNOWLEDGEMENT

The review of this paper by Dr. J.W. Hopmans of the Department of Land, Air and Water Resources, University of California is gratefully acknowledged.

REFERENCES

BASTIAANSSEN, W.G.M., 1991. Derivation of areal soil physical data from satellite measurements. Hydrological interactions between atmosphere, soil and vegetation. IAHS Publ. No. 204:95–105.

BASTIAANSSEN, W.G.M., 1993. Energy balance modelling from a hydrological perspective using operational earth observation satellite data. PhD. Thesis, Wageningen Agricultural University, Wageningen, in preparation.

BASTIAANSSEN, W.G.M. and R.A. ROEBELING, 1993. Analysis of land-surface exchange processes in two agricultural regions in Spain using Thematic Mapper Simulator data. In: H.J. Bolle, R.A. Feddes and J.D. Kalma (eds). Exchange processes at the land surface for a range of space and time scales. IAHS Publ. No. 212:407–416.

BASTIAANSSEN, W.G.M., D.H. HOEKMAN and R.A. ROEBELING, 1994. A methodology for the assessment of surface resistance and soil water storage variability at mesoscale based on remote sensing measurements (A case study with HAPEX-EFEDA data). IAHS Spec. Publ. No. 2 (Ed. H. Salz): 64 pp.

BELMANS, C., J.G. WESSELING and R.A. FEDDES, 1983. Simulation model of the water balance of a cropped soil: SWATRE. J.Hydrol., 63:271–286.

BEVEN, K., 1989. Changing ideas in hydrology – the case of physically based models. J.Hydrol., 105:157–172.

BEVEN, K.J., 1993. Estimating transport parameters at the grid scale: on the value of a single measurement. J.Hydrol. 143:109–123.

CHEN, Z.Q., M.L. KAVVAS and R.S. GOVINDARAJU, 1993. Upscaling of Richards' equation for soil moisture dynamics to be utilized in mesoscale atmospheric models. In: H.J. Bolle, R.A. Feddes and J.D. Kalma (eds), Exchange processes at the land surface for a range of space and time scales. IAHS. Publ. no. 212:125–132.

CHOUDHURY, B.J., 1991. Multispectral satellite data in the context of land surface heat balance. Rev. Geophys. 29(2):217–236.

DOBSON, M.C., F.T. ULABY, M.T. HALLIKAINEN and M.A. EL-RAYES, 1985. Microwave dielectric behaviour of wet soil – Part II: Dielectric mixing models. IEEE Trans. Geosci. Rem. Sens., GE-23(1):35–46.

FEDDES, R.A., P.J. KOWALIK and H. ZARADNY, 1978. Simulation of field water use and crop yield. Simulation monographs. PUDOC, Wageningen, Netherlands: 189 p.

FEDDES, R.A., P. KABAT, P.J.T. VAN BAKEL, J.J.B. BRONSWIJK and J. HALBERTSMA, 1988. Modelling soil water dynamics in the unsaturated zone -state of the art. J.Hydrol., 100:69–111.

FEDDES, R.A., G.H. DE ROOIJ, J.C. VAN DAM, P. KABAT, P. DROOGERS and J.N.M. STRICKER, 1993a. Estimation of regional effective soil hydraulic parameters by inverse modelling. In: D. Russo and G. Dagan (eds), Water Flow and Solute Transport in Soils: Springer Verlag, Berlin. Advanced Series in Agricult. Sciences. vol. 20:211–231.

FEDDES, R.A., M. MENENTI, P. KABAT and W.G.M. BASTIAANSSEN, 1993b. Is large-scale inverse modelling of unsaturated flow with areal average evaporation and surface soil moisture as estimated from remote sensing feasible? J.Hydrol., 143:125–152.

HOPMANS, J.W. and J.N.M. STRICKER, 1989. Stochastic analysis of soil water regime in a watershed. J.Hydrol., 105:57–84.

KIM, C.P., 1993. Coupling of hydrological models to climate models: aggregation and disaggregation. In: Scaling problems in hydrology (in Dutch). Commissie Hydrologisch Onderzoek, TNO, Delft, Rapporten en Nota's No. 31:77–93.

KOOL, J.B. and J.C. PARKER, 1987. Estimating soil hydraulic properties from transient flow experiments: SFIT User's Guide. Soil and Environmental Sciences, Virginia Polytechnic Institute and State University Blackburg, VA, 59 pp.

MANTOGLOU, A. and L.W. GELHAR, 1987. Effective hydraulic conductivities of transient unsaturated flow in stratified soils. Water Resour. Res., 23:57–68.

MENENTI, M., BASTIAANSSEN, W.G.M., VAN EICK, D., and ABD EL KARIM, M.H., 1989. Linear relationships between surface reflectance and temperature and their application to map actual evaporation of groundwater. In: V.V. Salomonson, L.S. Walter and C. Maetzler (eds), Remote Sensing of the Earth's Surface: Advances in Space Research, Pergamon, Oxford, pp. 165–176.

MISHRA, S., J.L. ZHU and J.C. PARKER, 1990. How effective are effective medium properties? In: ModelCARE 90: Calibration and reliability in Groundwater Modelling, Proc. Conf., The Hague, September 1990. IAHS Publication No. 195:521–528.

OH, Y., K. SARABANDI and F.T. ULABY, 1992. An empirical model and an inversion technique for radar scattering from bare soil surface. IEEE Trans. Geosci. and Rem. Sens., 30:370–381.

RAUPACH, M.R., 1991. Vegetation – atmosphere interaction in homogeneous and heterogeneous terrain: some implications of mixed-layer dynamics. Vegetatio 91:105–120.

SCHMUGGE, T.J. and F. BECKER, 1991. Remote sensing observations for the monitoring of land surface fluxes and water budgets. In: T.J. Schmugge and J.C. Andre (eds), Land Surface Evaporation. Measurement and Parametrization. Springer Verlag, Berlin: 227–347.

SHAMIR, U., 1992. Personal communication.

SHI, J.V., J.V. SOARES, L. HESS, E.T. ENGMAN and J.J. VAN ZIJL, 1991. Soil moisture measurements from airborne SAR. Proc. of the 3rd Airborne Synthetic Aperture Radar (AIRSAR), Pasadena, JP-publication, 91–30:68–77.

STRICKER, J.N.M., C.P. KIM, R.A. FEDDES, J.C. VAN DAM, P. DROOGERS and G.H. DE ROOIJ, 1993. The terrestrial hydrological cycle. In: F. Raschke and D. Jacob (eds), Energy and Water Cycles in the Climate System. NATO ASI Series, Springer Verlag, Berlin, Heidelberg, vol. I5:419–444.

ULABY, F.T. and C. ELACHI (eds), 1990. Radar polarimetry for geoscience applications. Artech House, London: 364 pp.

VAN GENUCHTEN, M.Th., 1980. A closed-form equation for predicting the hydraulic conductivity of unsaturated soils. Soil Sci. Soc. Am. J44:892–898.

WANG, J.R. and T.J. SCHMUGGE, 1980. An empirical model for the complex dielectric permittivity of soil as a function of water content. IEEE Trans. on Geosci. Rem. Sens., 18:288–295.

The importance of landscape position in scaling SVAT models to catchment scale hydroecological prediction

T.J. HATTON, W.R. DAWES and R.A. VERTESSY

CSIRO Division of Water Resources,

GPO Box 1666, Canberra, ACT 2601

Australia

ABSTRACT Process-based models of soil-vegetation-atmosphere interactions developed for small plots (points) define vertical transfers of water and energy. One can attempt to scale to larger heterogeneous land units by disaggregating the landscape into a set of elements and applying a vertical SVAT model independently to each element (Running et al., 1989; Pierce et al., 1992). Such applications fail to consider lateral transfers. A distributed parameter, three-dimensional SVAT (Topog-IRM) developed by the CSIRO Division of Water Resources (O'Loughlin, 1990; Hatton and Dawes, 1991; Hatton et al., 1992) is used to examine the importance of lateral transfers of water for prediction of water balance components at the small catchment scale.

Simulations are used to contrast the predicted water balances from a SVAT model with and without considerations of lateral subsurface and overland flow in complex terrain. Components of the catchment water balance are shown to scale linearly except in those cases where transient perched water tables develop in landscapes with sufficient slope and hydraulic conductivity to redistribute water effectively via subsurface lateral flow. In such cases, the prediction of catchment yield and the spatial pattern of soil moisture requires the explicit treatment of lateral transfers.

1. INTRODUCTION

The most widely-used soil-vegetation-atmosphere (SVAT) models calculate the surface energy and water (and carbon) balances in the vertical dimension only (e.g. Running and Coughlan, 1988; Wang and Jarvis, 1990). Extension of plot-scale SVAT models to larger scales (Running et al., 1989; Pierce et al., 1992) disaggregate the simulated region into a regular grid of elements, each with a particular loading of radiation and rainfall. Vertical SVAT models (Running and Coughlan, 1988; Hatton et al., 1992) are used to simulate the water and energy balances in each element. Band et al. (1991) described a framework (RHESSYS) for modelling the water balance over complex terrain. In their work, however, the spatial expression of moisture resulted from the way in which the landscape was disaggregated into homogeneous land units as opposed to the explicit modelling of lateral water redistribution. They assumed that lateral transfers of water and energy are not significant to the landscape water balance. This assumption can be tested through simulations of catchment behaviour with and without lateral transfers.

The small catchment scale (10–10000 ha) is of particular significance in most landscape hydroecological determinations, including soil erosion, dryland salinisation, flood generation, water yield and vegetation dynamics. Additionally, it is significant in that it is the scale at which most land is managed. It is also a scale at which the lateral redistribution of water may be of critical importance to landscape water balance.

Several physically-based, three-dimensional models have been developed to address processes at this scale. WATSIM (Aston and Dunin, 1980) redistributed overland flow across arbitrary landscape units in a simulated catchment and demonstrated a marginal improvement in the prediction of water yield as opposed to simulating the land unit as a lumped system.

Both the SHE (Abbott et al. 1987a, b) and Topog (O'Loughlin, 1990; Hatton and Dawes, 1991; Hatton et al., 1992) models are physically-based, deterministic, distributed parameter catchment models capable of simulating the lateral transfers of water by the processes of overland flow, subsurface (saturated) lateral flow and a linkage between an unsaturated zone (vertical) model and a groundwater model.

Importantly, neither model addresses lateral transfers of energy (advection) from one landscape element to another. This is because our understanding of the interaction between the planetary boundary layer development over heterogeneous land surfaces in uneven terrain is limited (e.g., Raupach, 1991). In this paper we explore only the importance of the redistribution of liquid water in scaling SVAT models to small catchments.

We apply a model, Topog-IRM (*Integrated Rate Methodology*) to address the importance of water redistribution in scaling SVAT models to small catchments with respect to the following responses: (a) total catchment outflow (yield), (b) total catchment evaporation, and (c) the spatial pattern of soil moisture. We examine these responses at extremes of annual rainfall. Finally, we apply the plant growth simulation capabilities of Topog-IRM to explore some of the hydroecological consequences to scaling SVAT models.

2. METHODS

2.1 Description of Topog-IRM

Topog-IRM is composed of a basic topological kernel (Topog) and a model (IRM) which simulates the water, energy, carbon, and nitrogen balances. These solutions are modelled at a nominal time step of one day, except when numerical convergence criteria demand finer temporal resolution.

The first requirement for catchment behaviour modelling is the representation of the land surface in three dimensions. Topog (O'Loughlin et al., 1989; O'Loughlin, 1990) begins with a digital elevation model (DEM) of the area to be modelled, and applies a set of topological rules to (a) define a catchment boundary, and (b) calculate a network of landscape elements defined by lines of minimum distance between adjacent contours. The network of elements is arranged around critical topographic features such as peaks, saddles, ridges, drainage lines, and stream confluences; the result is a series of adjacent flow strips which diverge or converge according to local terrain.

The second requirement for landscape scale modelling is

expressing the spatial distribution of parameters. Within the Topog framework, the attributes slope, aspect, extraterrestrial radiation, and specific catchment area are calculated for each element. The spatial patterns for soil, rainfall, vegetation type, biomass, and plant nutrients can be superimposed on the element network.

Infiltration of net rainfall and the vertical redistribution of water in the soil are accomplished through a finite difference numerical solution of the Richards' equation (Ross, 1990) for each landscape element. Matrix flow alone is considered for water redistribution in the vertical (preferred pathway flow and hysteresis are not modelled). Overland flow, whether resulting from infiltration excess or surface saturation, is treated explicitly. Following the development of a water table, water is moved laterally according to the saturated hydraulic conductivity and the local hydraulic head. Linkage to the deep groundwater system (if present) can also be modelled, but is not considered here.

Each soil type in the catchment is ascribed a set of hydraulic attributes. A table defining the relationships among water potential, volumetric water content and hydraulic conductivity is obtained using the analytical solution of Broadbridge and White (1988). This table is referenced by the infiltration submodel described above.

The largest term in the annual water balance for most land surfaces, after gross rainfall, is evaporation. Thus, the partitioning of the surface energy balance into latent and sensible heat is crucial to models of catchment behaviour. The direct input to the evaporation module is daily meteorological information on maximum and minimum temperatures, vapour pressure deficit, precipitation, and direct and diffuse solar radiation incident on the horizontal plane. Given these inputs, the evaporation module calculates daily values for evaporation of water intercepted by the canopy, transpiration and soil evaporation for each element. Each of these processes is related to the extent of the canopy leaf area index (LAI).

The energy available for evaporation is limited to the amount of net radiant energy received by each element, as modified by the average daily vapour pressure deficit. The horizontal advection of sensible heat is not treated; the energy balance of each element is thus independent of its neighbours. Direct radiation is distributed throughout the catchment according to the aspect and slope of each element and the time of year; diffuse radiation is modified by slope alone. Canopy average photosynthetically active radiation (PAR) is calculated by the Beer-Lambert Law as in Running and Coughlan (1988). Net radiation is resolved as the sum of total absorbed shortwave and net longwave; the latter is calculated as in Jury and Tanner (1975).

Rainfall interception is scaled by the daily effective canopy storage per unit LAI. The term 'effective' is used to accommodate the fact that over a day, the total storage capacity exceeds the instantaneous value (Dunin et al., 1988). The energy required to evaporate intercepted water is deducted from that available for transpiration.

The amount of net radiant energy received at the soil surface sets an upper boundary condition for evaporation at the uppermost node in the unsaturated zone model described above. Following Gardner (1959), the amount of water evaporated is limited initially by energy and, as the surface dries, by the hydraulic conductivity of the soil. Soil heat flux is assumed to approach zero on a daily time step.

The key controls on transpiration are the two resistance terms. For very rough surfaces like forests, the aerodynamic resistance may be treated as small and effectively independent of windspeed and atmospheric stability. The surface resistance, however, will vary widely in time, and is largely a function of the stomatal control of water vapour transport in the canopy. This control is in turn dictated by the availabilities of light, water, carbon dioxide, and nutrients, as modified by temperature and vapour pressure deficits (Ball et al., 1987; Lynn and Carlson, 1990).

Topog-IRM implements the empirical model of stomatal conductance (g_c) of Ball et al. (1987) and Collatz et al. (1991), which relates conductance to carbon assimilation rate (A), humidity, and the concentration of CO_2. The estimation of A is discussed in the following section. Surface resistance is assumed to approach zero when the canopy is wet.

Once an estimate of daily transpiration is obtained, that amount of water must be extracted from the soil profile of each landscape element. This total is apportioned vertically on the basis of the relative rooting density and the water potential with depth. Rooting density is assumed to fall off exponentially to zero at the maximum rooting depth as in Ritchie et al. (1986).

The effects of the various factors which influence carbon dioxide assimilation are widely reported (see, for example, Wong et al., 1978, 1985a, 1985b). A recent advance in the mathematics associated with saturation rate kinetics (Wu et al., 1991) extended the single factor hyperbolic velocity (Michaelis-Menten) curve into a framework for integrating the effects of n factors on the rate of any process in which the primary responses are hyperbolic. By applying the multiple factor equations of Wu et al. (1992), we model carbon assimilation (A) by saturation rate kinetics to a four factor model in which assimilation is determined by the availabilities of light, water, nitrogen, and CO_2, as modified by temperature and the vapour pressure deficit. The estimate of A is then used in the equation for stomatal conductance. Topog-IRM uses the daily estimate of gross assimilation by each landscape element to calculate a total plant carbon balance, accounting for respiration and allocation. More

detailed descriptions of the governing equations and parameterisations for plant growth, and how these equations are linked to the nitrogen cycle, are in Hatton et al. (1992).

Topog-IRM outputs a range of hydroecological responses. For each landscape element, the water content and potential at each depth node i are calculated, as well as the mean water content for the entire soil column. Daily transpiration, rainfall interception, soil evaporation, deep drainage, lateral subsurface flow, and infiltration excess are available for each element. Carbon components in leaf, stem, and root compartments, total mineralisable nitrogen, and total available nitrogen are also calculated for each element.

To contrast the effects of scaling SVATS to small catchments with and without lateral drainage, the fully linked version (hereafter referred to as Model A) was modified such that lateral subsurface flow was disabled (Model B). Thus, in either model, water in excess of soil water holding capacity for any element was transferred to the catchment outlet (and thus treated as yield) within the same time step. Under Model B, however, no water is yielded by subsurface lateral movement. Model B, therefore, encompasses only the first level domain (Becker and Nemec, 1987), while Model A considers both the first and second level domains (vertical as well as lateral transfers).

2.2 Catchment simulations with (Model A) and without (Model B) lateral flow

Three sets of simulations were carried out. A summary of these simulations appears in Table 1. All simulations use a DEM and climatic data from a 32 ha catchment in Victoria, Australia known as Myrtle II. This catchment forms part of Melbourne's municipal water supply, and is covered by an even-aged forest dominated by mountain ash (*Eucalyptus regnans*). A fuller description of this catchment, its history and its importance is in Vertessy et al. (1991).

Simulation set 1 of Table 1 compared outputs from Models A and B for the catchment given a spatially uniform soil depth of 2.0 m and a uniform LAI of 3.0, with vegetation consisting of a forest of *E. regnans*. Simulations were run with climatic data for a dry and a wet time series of 600 days (with total precipitations of 969 and 1935 mm, respectively, uniformly distributed across the catchment). Plant growth was not simulated. Initial state variable and parameter values appear in Table 2.

Outputs from these simulations show water balance components with time on (a) a total catchment basis, and (b) for a near-ridge element and a valley bottom element. These latter landscape units were chosen to ensure a similar aspect and slope. Additionally, the spatial patterns in relative moisture content on year day 50 (the 600th day of simulation) were also obtained.

Table 1. *Design of catchment simulations comparing Models A and B.*

	Simulation sets		
	1	2	3
Soil depth (m)	2.0	5.0	2.0
Vegetation type	forest	forest	pasture
LAI	3.0	3.5–6.0	1.0
Plant growth	no	no	yes
Rainfall (mm)	969 and 1935	1133	1935
Length (days)	600	365	600

Simulation set 2 (Table 1) compared hydrographs predicted by Models A and B with an actual hydrograph for the Myrtle II catchment for the year 1972. In this case, the soil was again uniform but with the more realistic depth of 5 m and with parameters for the Broadbridge and White (1988) soil hydraulic model as described in Vertessy et al. (1991). In this case, LAI of the mountain ash was distributed in space such that an area defined around the valley bottoms supported an LAI of 6.0, while the rest of the catchment had an LAI of 3.5; this distribution reflects the actual pattern of LAI in Myrtle II. Other parameters remained the same as in Table 2. Again, no plant growth was simulated.

In simulation set 3 (Table 1), the effects of the spatial redistribution of water on the landscape vegetation response were demonstrated (with Model A) by utilising the plant growth modelling capability of TOPOG-IRM. Simulations were performed using a vegetation cover of C3 (cool season) pasture at an initial LAI of 1.0 to emphasise the importance of the spatial variability of water on site productivity. Soil hydraulic properties were as in Table 1. Nitrogen was made non-limiting to plant growth.

3. RESULTS

Water balance components from simulation set 1 (Table 1) appear in Table 3. The most obvious feature of the results is that those components which are modelled in the vertical domain (rainfall interception, soil evaporation, transpiration) are similar between Models A and B. The slight depression of total transpiration in the *wet year* simulation of Model A is due to the drainage of water away from the upper catchment, thus slightly reducing its availability to plants relative to Model B.

In the case of the *dry year* simulations, neither water tables nor overland flow developed; as a result, no yield was predicted by either model and the system was adequately modelled in the vertical domain. In the *wet* simulation,

Table 2. *Initial state variables and parameter values for Topog-IRM (Model A) as applied to a catchment covered by a forest of* Eucalyptus regnans.

Initial values	Description	Unit
State variables		
−0.1	Soil water potential	MPa
0.50	Leaf carbon	$kg\ m^{-2}$
0	Stem carbon	$kg\ m^{-2}$
0.50	Root carbon	$kg\ m^{-2}$
Parameter variables		
2.00	Saturated hydraulic conductivity	$m\ d^{-1}$
0.35	Saturated moisture (vol/vol)	
0.15	Air-dry moisture (vol/vol)	
0.20	Capillary length scale	m
1.05	Soil structural parameter	
2	Specific leaf area	$m^2\ kg^{-1}$
−0.42	Light extinction coefficient	
0.0003	Rainfall interception coefficient	$m\ LAI^{-1}\ d^{-1}$
12	Aerodynamic resistance	$s\ m^{-1}$
0.15	Canopy albedo	
0.15	Soil albedo	
9.9	Slope in conductance model	
0.01	Intercept in conductance model	$mol\ m^{-2}\ s^{-1}$
−1.65	Minimum available water potential	MPa
15.0	Maximum assimilation rate	$mol\ m^{-2}\ s^{-1}$
2.13	Weighting factor for water	
0.11	Weighting factor for nitrogen	
1.42	Weighting factor for CO_2	
1000	Light saturation point	$umol\ m^{-2}\ s^{-1}$
295	Upper half-saturation temperature for photosynthesis	K
275	Lower half-saturation temperature for photosynthesis	K
0.0003	Leaf respiration coefficient	$kg\ kg^{-1}\ C^{-1}\ d^{-1}$
0.0007	Stem respiration coefficient	$kg\ kg^{-1}\ C^{-1}\ d^{-1}$
0.0002	Root respiration coefficient	$kg\ kg^{-1}\ C^{-1}\ d^{-1}$
0.35	Leaf growth respiration	$kg\ kg^{-1}$
0.35	Root growth respiration	$kg\ kg^{-1}$
0.022	Maximum leaf nitrogen	$kg\ N\ kg\ C^{-1}$
0.011	Minimum leaf nitrogen	$kg\ N\ kg\ C^{-1}$
0.001	Stem nitrogen	$kg\ N\ kg\ C^{-1}$
0.008	Root nitrogen	$kg\ N\ kg\ C^{-1}$

Table 3. *Simulated water balance components (mm) for a small catchment with (Model A) and without (Model B) lateral connectivity in water flows, for a wet and a dry climate (1935 and 969 mm, respectively).*

	Wet		Dry	
	A	B	A	B
Rainfall	1935	1935	969	969
Interception	205	205	188	188
Soil evaporation	390	390	269	269
Transpiration	848	858	499	499
Storage change	111	244	14	14
Total discharge	381	238	0	0

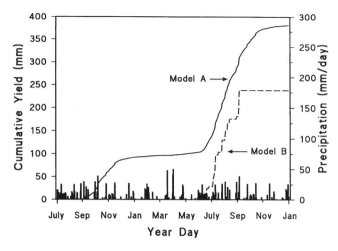

Fig. 1 Comparison of simulated cumulative discharge yields from Models A and B with a high rainfall input.

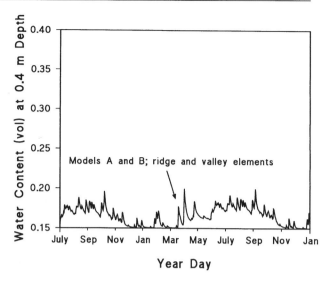

Fig. 2 Simulated volumetric water contents at 0.4 m depth from Models A and B with a low rainfall input. Because no water table forms, the predicted responses from the two models are identical regardless of landscape position.

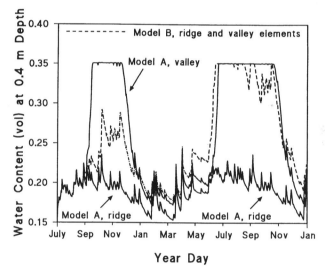

Fig. 3 Simulated volumetric water contents at 0.4 m depth from Models A and B with a high rainfall input. In this case, formation of perched water tables under Model A leads to subsurface flow and the redistribution of water from the upper portions of the catchments into the valley. Thus, the simulated ridge element remains relatively drier than the valley element, while the responses of the elements under Model B are identical.

however, substantial drainage via subsurface flow was predicted by Model A. As a result, 60 per cent more discharge occurred relative to Model B simply by taking into account lateral redistributions.

The temporal pattern of this discharge is depicted in Fig. 1. With Model A, discharge yield began earlier in the simulation relative to Model B, and continued through the spring. The simplistic treatment of yield by Model B is reflected in the on/off behaviour of yield; the landscape elements only contribute to yield when they are fully saturated.

Examination of the behaviour of a pair of (otherwise) similar ridge and valley elements shows the importance of considering lateral flow. In the dry year, the water contents of the elements at an arbitrary depth (0.4 m) were identical all year (Fig. 2). This is expected, given the lack of any water table development and the fact that each element received the same amount of radiation over the simulation.

In the wet simulation, the water contents at 0.4 m were quite different between the ridge and valley elements (Fig. 3), due to the redistribution of water via lateral subsurface slow. Water contents under Model B were identical for these elements, and were midway between the values under Model A. More subtle differences between the models were evident in the valley element responses; in Model B, the water content varied greatly during the two winter periods. This reflected the (vertical) balance between rainfall and evapotranspiration in that element alone. Model A, however, was constantly draining water into this lower element, and thus it tended to stay saturated.

The spatial pattern of percentage saturation across the catchment on the 600th day of simulation for the wet period resulting from Model A (Fig. 4) shows a realistic distribution of moisture in space for that time of year. Valley bottoms are wet, while upper portions of the catchments are drier. The moisture pattern for the same conditions from Model B is

quite uniform, ranging only between 0.62 and 0.66 per cent saturation (Fig. 5). This pattern is explained by the differential loading of radiation, primarily as a function of slope (Fig. 6).

Simulation set 2 (Table 1) tested predictions by Models A and B against an observed hydrograph from the Myrtle II catchment. As in the first set of simulations, there were only trivial differences in simulated interception, soil evaporation,

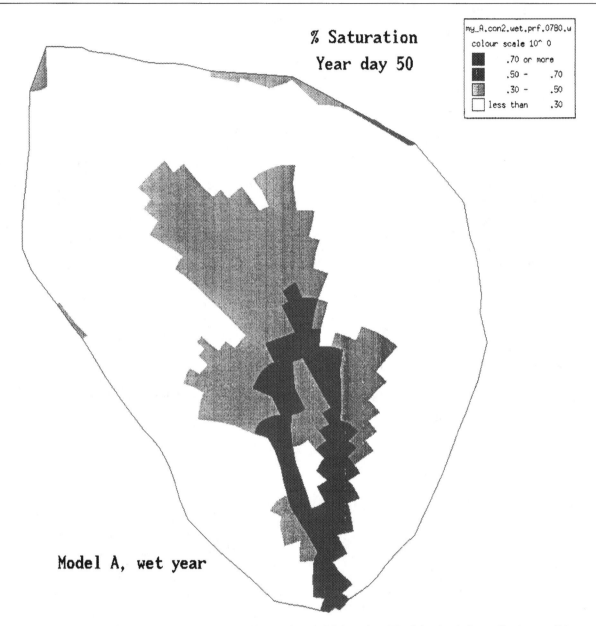

Fig. 4 The distribution of soil wetness (percentage saturation) under Model A on day 600 of the simulation, reflecting conditions expected in early (southern) summer.

and transpiration (Table 4). Once again, however, explicit treatment of lateral subsurface flow led to a very large difference in predicted yield. In this case, Model B yielded only a small fraction of the observed catchment discharge yield, while Model A was only about 2 per cent in error. Importantly, Model A also successfully reproduced the daily hydrograph (Fig. 7), while Model B only yielded water over a few days in early spring.

Simulation set 3 (Table 1) was aimed at demonstrating the ecological consequences to modelling water in the horizontal as well as the vertical. In this case, the pasture growth modelling capabilities of Topog-IRM were enabled and parameterised for a winter-dominant pasture with an initial

LAI of 1.0. By the 600th day of simulation (mid-summer), the predicted pattern of LAI reflected the expected availability of water across the catchment (Fig. 8). Model B could not reproduce this pattern, as evidenced by the predicted distribution of water in Fig. 5.

4. DISCUSSION

The nature of the soil-plant-atmosphere continuum is such that the dominant processes, at least near the surface, are in the vertical domain. In addition, the facts that a soil has a finite capacity to hold water, and that the atmosphere has a

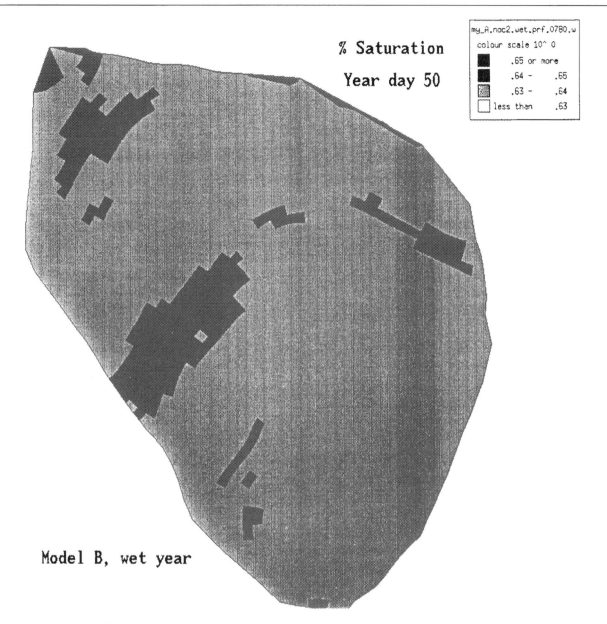

% Saturation

Year day 50

my_A.noc2.wet.prf.0780.w
colour scale 10^ 0
■ .65 or more
■ .64 – .65
▨ .63 – .64
☐ less than .63

Model B, wet year

Fig. 5 The distribution of soil wetness (percentage saturation) under Model B on day 600 of the simulation, reflecting conditions expected in early (southern) summer. In this case, the omission of lateral flow leads to an essentially uniform distribution of moisture across the catchment (note range of colour scale). Only small differences in radiation interception induce variability.

finite capacity for evaporation, make strictly vertical treatments of the surface water balance inherently conservative. That is, simulations of the soil water balance are naturally bounded between the air-dry and saturation limits of soil water storage, the capacity of the canopy to intercept rainfall, and the potential evaporation rate. It is therefore not too suprising that a vertical approach (i.e., Model B) to modelling the components of the annual water balance of a plot works reasonably well for soil evaporation, transpiration and rainfall interception – all processes subject to these limits. On a catchment basis, the areal average of these responses may or may not be unbiased, depending on the internal variance of the landscape attributes.

Catchment response in terms of water yield is not wholly subject to these constraints. Given the development of a perched water table and a hydraulic gradient, lateral subsurface flow is generated. This process redistributes water within the catchment, and delivers water at some rate to the outlet. It also establishes the source area for stormflow and creates a diversity of plant community niches. Models of catchment response limited to the vertical domain do not reproduce this behaviour. Thus, naive models such as Model B in this paper 'suspend' water in landscape positions which in nature would drain downslope. The general prediction of catchment yield therefore requires some statistical treatment of yield resulting from the expected water balance in the vertical domain,

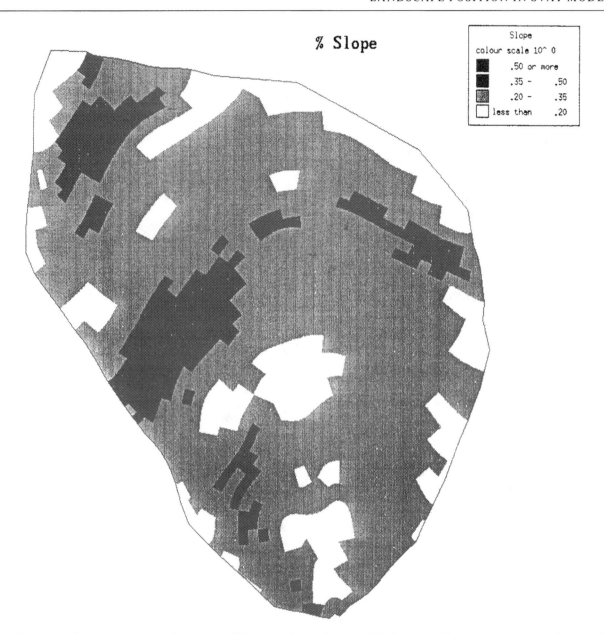

Fig. 6 Percentage slope across simulated catchment. This pattern is largely responsible for the small spatial variance in moisture under Model B through its effects on radiation loading.

or a process-based treatment incorporating lateral transfers (i.e., Model A). Band et al. (1991) also concluded that regional SVAT models need to incorporate water movement in three dimensions. Such processes may be expected to dominate the flood hydrograph most commonly in humid regions with permeable soils on steep hillslopes (Dunne, 1983).

There are, however, cases in which one can expect a vertical model to scale linearly. Where the evaporation rate approximates rainfall and thus no water table ever develops, a simple vertical SVAT is sufficient to predict no yield in catchments not subject to any regional water table (closed). Similarly, as the hydraulic conductivity or the slope of the

catchment decreases, the approximation of yield by a vertical model should inprove.

It is unreasonable to disaggregate a landscape at the scale of 10^1–10^4 ha into a set of elements which are modelled as spatially independent units and expect to reproduce the daily hydrograph from a series of climatic inputs. This paper demonstrates both the inadequacy of such a simplistic treatment as well as the power of a physically-based process model (Topog-IRM) in simulating this response from a catchment with complex terrain.

Models of catchment behaviour operating solely in the vertical domain can only express the spatial variability of moisture as a function of the spatial variation in radiation

Table 4. *Simulated water balance components (mm) for the Myrtle II catchment with (Model A) and without (Model B) lateral connectivity and actual water yield for 1972.*

	Model A	Model B	Actual
Rainfall	1133	1133	1133
Interception	238	238	
Soil evaporation	136	151	
Transpiration	579	586	
Storage change	−453	127	
Total discharge	633	31	616

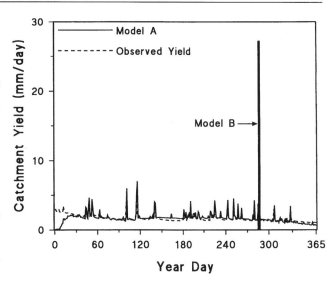

Fig. 7 Simulated daily hydrographs under Models A and B for a one year simulation for the Myrtle II catchment with the observed hydrograph. Model A successfully reproduces the observed behaviour through its consideration of water redistribution. Model B only generates yield when elements are fully saturated.

(assuming spatially uniform rainfall and soil). This effect is of a minor order when compared to moisture redistribution by lateral flow. Further, the horizontal advection of energy, not treated in these simulations, will tend to blur these spatial differences in evaporation at these scales.

The consequences for hydroecological prediction in incorporating the lateral domain are considerable. Were we to infer the productivity or ecological niche of a given species or plant community from Model B, we would expect the entire catchment to have some average moisture availability with a relatively low spatial variance. In fact, we normally observe in nature a large spatial variance in soil moisture in complex, hilly terrain. Associated with this spatial pattern are gradients in primary productivity and plant community composition. We demonstrated the expected spatial expression of the former in this paper by applying the plant growth capabilities of Topog-IRM. This simulation reflected the importance of extending SVAT models to the horizontal domain in scaling hydroecological prediction from the plot to catchment scales.

REFERENCES

ABBOTT, M.B., BATHHURST, J.C., CUNGE, J.A., O'CONNELL, P.E. and RASMUSSEN, J. (1986a) An introduction to the European Hydrological System – Système Hydrologique Européen, 'SHE.' I. History and philosophy of a physically-based, distributed parameter modelling system. J. Hydrol. 87, 45–59.

ABBOTT, M.B., BATHHURST, J.C., CUNGE, J.A., O'CONNELL, P.E. and RASMUSSEN, J. (1986b) An introduction to the European Hydrological System – Système Hydrologique Européen, 'SHE.' II. Structure of a physically-based, distributed modelling system. J. Hydrol. 87, 61–77.

ASTON, A.R. and DUNIN, F.X. (1980) The prediction of water yield from a 5 ha experimental catchment, Krawarre, N.S.W. Aust. J. Soil Res. 18, 149–62.

BALL, J.T., WOODROW, I.E. and BERRY, J.A. (1987) A model predicting stomatal conductances and its contribution to the control of photosynthesis under different environmental conditions. In: Progress in Photosynthesis Research Vol. IV. (J. Biggins, ed) pp. 221–224. Martinus Nijhoff, Dordrecht.

BAND, L.E., PETERSON, D.L., RUNNING, S.W., COUGHLAN, J.C., LAMMERS, R., DUNGAN, J. and NEMANNI, R. (1991)

Forest ecosystem processes at the watershed scale: basis for distribution simulation. Ecol. Model. 56, 171–96.

BECKER, A. and NEMEC, J. (1987) Macro-scale hydrologic models in support of climate research. In: The Influence of Climate Change and Climatic Variability on the Hydrologic Regime and Water Resources. Proceedings of the Vancouver Symposium, Aug. 1987. IAHS Publication No. 168.

BROADBRIDGE, P. and WHITE, I. (1988) Constant rate rainfall infiltration: a versatile nonlinear model. I. Analytic solution. Water Resources Res. 24, 145–54.

COLLATZ, G.J., BALL, J.T., GRIVET, C. and BERRY, J.A. (1991) Physiological and environmental regulation of stomatal conductance, photosynthesis and transpiration: a model that includes a laminar boundary layer. Agric. For. Meteor. 54, 107–36.

DUNIN, F.X., O'LOUGHLIN, E.M., and REYENGA, W. (1988) Interception loss from a Eucalypt forest: lysimeter determination of hourly rates for long term evaluation. Hydrol. Process. 2, 315–29.

DUNNE, T. (1983) The relation of field studies and modeling in the prediction of storm runoff. J. Hydrol. 65, 25–48.

GARDNER, W.R. (1959) Solutions of the flow equations for the drying of soils and other porous media. Soil Science Soc. Amer. J. 23, 183–87.

HATTON, T.J. and DAWES, W.R. (1991) The impact of tree planting in the Murray-Darling Basin: The use of the TOPOG-IRM hydroecological model in targeting tree planting sites in catchments. CSIRO Division of Water Resources Tech. Memo. 91/14, Canberra, Australia.

HATTON, T.J., DAWES, W.R., WALKER, J. and DUNIN, F.X. (1992) Simulations of hydroecological responses to elevated CO_2 at the catchment scale. Aust. J. Bot. 40, 679–696.

HATTON, T.J., PIERCE, L.L. and WALKER, J. (1993) Ecohydrological changes in the Murray-Darling Basin: II. Development and tests of a water balance model. J. Appl. Ecol. 30, 274–282.

JURY, W.A. and TANNER, C.B. (1975) Advection modification to the Priestley and Taylor evapotranspiration formula. Agron. J. 67, 840–42.

LYNN, B.H. and CARLSON, T.N. (1990) A stomatal resistance model illustrating plant vs. external control of transpiration. Agric. For. Meteor. 52, 5–43.

O'LOUGHLIN, E.M. (1990) Modelling soil water status in complex terrain. Agric. Forest Meteor. 50, 23–38.

O'LOUGHLIN, E.M., SHORT, D.L. and DAWES, W.R. (1989) Modelling the hydrological response of catchments to landuse change. In Hydrology and Water Resources Symposium, Institution

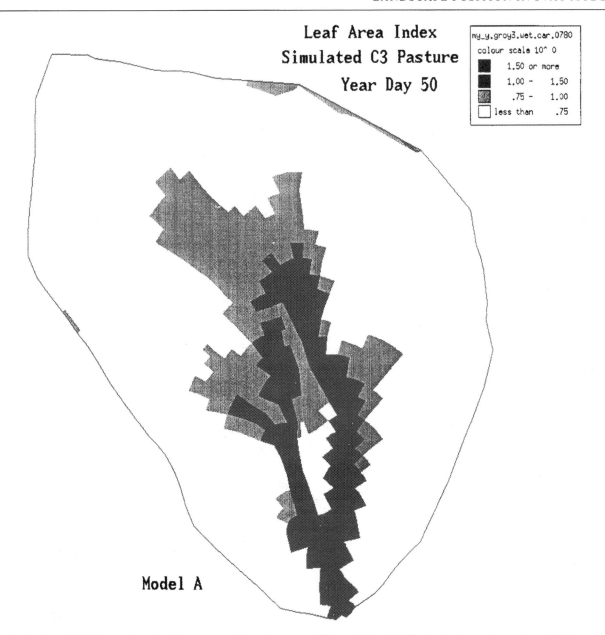

Fig. 8 Simulated LAI for a winter-dominant pasture in early summer. The spatial variability in plant growth covaries with the distribution of water availability.

of Engineers, Australia, Publication No. 89/19. Christchurch, New Zealand pp. 335–40.

PIERCE, L.L., WALKER, J., DOWLING, T.I., MCVICAR, T.R., HATTON, T.J., RUNNING, S.W. and COUGHLAN, J.C. (1993) Ecohydrological changes in the Murray-Darling Basin: III. A simulation of regional hydrological change. J. Appl. Ecol. 30, 283–294.

RAUPACH, M.R. (1991) Vegetation-atmosphere interaction in homogeneous and heterogeneous terrain: some implications of mixed-layer dynamics. Vegetatio 91, 105–20.

RITCHIE, J.T., KINIRY, J.R., JONES, C.A., and DYKE, P.T. (1986) Model inputs. In: CERES-MAIZE: A Simulation Model of Maize Growth and Development (C.A. Jones and J.R. Kiniry, eds). Texas A&M Press, College Station, pp. 37–48.

ROSS, P.J. (1990) Efficient numerical methods for infiltration using Richards' equation. Water Resources Res. 26, 279–90.

RUNNING, S.W. and COUGHLAN, J.C. (1988) A general model of forest ecosystem processes for regional applications. I. Hydrologic balance, canopy gas exchange and primary production processes. Ecol. Model. 42, 125–154.

RUNNING, S.W., NEMANI, R.R., PETERSON, D.L., BAND, L.E., POTTS, D.F., PIERCE, L.L. and SPANNER, M.A. (1989) Mapping regional forest evapotranspiration and photosynthesis by coupling satellite data with ecosystem simulation. Ecology 70, 1090–101.

VERTESSY, R.A., HATTON, T.J., O'LOUGHLIN, E.M. and BROPHY, J.H. (1991) Water balance simulation in the Melbourne water supply area using a terrain analysis-based catchment model. CSIRO Division of Water Resources Consultancy Report 91/32, Canberra, Australia.

WANG, Y.P. and JARVIS, P.G. (1990) Influence of crown structural properties on PAR absorption, photosynthesis, and transpiration in Sitka spruce: application of a model (MAESTRO). Tree Physiol. 7, 297–316.

WONG, S.C., COWAN, I.R. and FARQUHAR, G.D. (1978) Leaf conductance in relation to assimilation in Eucalyptus pauciflora Sieb. ex Spreng. Influence of irradiance and partial pressure of carbon dioxide. Plant Physiol. 62, 670–74.

WONG, S.C., COWAN, I.R. and FARQUHAR, G.D. (1985a) Leaf conductance in relation to rate of CO_2 assimilation. I. Influence of nitrogen nutrition, phosphorous nutrition, photon flux density, and ambient partial pressure of CO_2 during ontogeny. Plant Physiol. 78, 821–25.

WONG, S.C., COWAN, I.R. and FARQUHAR, G.D. (1985b) Leaf conductance in relation to rate of CO_2 assimilation. II. Effects of short-term exposures to different photon flux densities. Plant Physiol. 78, 826–29.

WU, H., RYKIEL. E.J., SHARPE, P.J., WALKER, J., HATTON, T.J., SPENCE, R.D. and LI, Y. (1994) Multi-factor growth rate modelling using an integrated rate methodology. Ecol. Model. 73, 97–116.

The influence of subgrid-scale spatial variability on precipitation and soil moisture in an atmospheric GCM

D. ENTEKHABI

48–331 Ralph M. Parsons Laboratory,

Department of Civil and Environmental Engineering,

Massachusetts Institute of Technology,

Cambridge, Massachusetts 02139 USA

ABSTRACT A new land surface hydrological parameterization for atmospheric General Circulation Models is introduced. The model incorporates physically-based relations for the partitioning of atmospheric energy and moisture forcing. Using statistical-dynamical derived-distribution techniques, closed-form and computationally-efficient expressions are developed for the inclusion of subgrid-scale spatial variability into parameterizations for atmospheric models. Numerical experiments with a General Circulation model show improved water balance estimates.

1. INTRODUCTION

Atmospheric General Circulation Models (GCMs) integrate the relevant conservation and state equations over grids (in the case of models using finite-difference schemes) or up to wave-numbers (in the case of models using spectral-solution techniques) that explicitly resolve processes whose dynamics occur at spatial scales on the order of hundreds of kilometres or larger. All physical processes whose characteristic scales are smaller are therefore either only implicitly represented or parameterized within the GCM. Radiative heating and cooling, turbulent diffusion, local convection, water phase transitions and hydrological processes are all examples of processes that are critical in the forcing of atmospheric motions, yet they are only parameterized in GCMs. The capability of GCMs of reproducing regional features of the climatic system and of capturing the significant feedbacks and interactions is thus partially dependent on the realism with which these physical processes are represented within the numerical model. In this paper the focus is on one of these processes, namely the land surface hydrology. As with other *physics* components in GCMs, the land surface hydrology is designed to capture the effective response of a system that depends on processes at a considerably finer scale than that of the GCM *dynamics*.

In this paper, a brief summary of the current approaches to parameterizing the soil and vegetation response to atmospheric forcing is given. The focus is on the manner in which the schemes partition the incident precipitation and radiative forcing between loss and storage terms. In this context, a new approach to dealing with hydrological processes that includes the effects of subgrid-scale spatial variability is introduced. The model is implemented into a coarse resolution GCM and the model climate is validated against the available observations of hydrological flux at various scales. In the final concluding section, a summary is given of the various ways in which the land surface hydrology may impose a strong influence on GCMs and their model climates.

2. PARAMETERIZATION OF LAND SURFACE HYDROLOGY

The land surface hydrology parameterization in GCMs performs various tasks in partitioning the atmospheric precipitation and radiative forcing. The storage and loss of heat and moisture by the plant canopy and soil column is chief among the tasks. The momentum transfer and the development of an atmospheric boundary layer with diffusive and turbulent-eddy transfer mechanisms is also an important function of the surface specification in GCMs.

In this brief overview, the focus is on the partitioning of incident rainfall into runoff and infiltration and the reduction of the atmospheric available energy forcing (in terms of energy-limited potential evaporation, E_p) into actual evaporation. There are numerous other important processes (such as transfer efficiencies in the surface layer, snow and ice processes, etc.) which will not receive direct treatment in this brief overview.

The runoff loss is a fraction R of the precipitation forcing; a fraction $(1 - R)$ of the incident precipitation is partitioned to storage at any time. The role of the land surface hydrology is to characterize the partitioning ratio R in terms of surface conditions. Clearly the state of the near-surface soil, its textural composition, hydraulic conductivity, level of saturation, etc. all influence the value of R. These characteristics are all spatially variable at scales of few tens of metres or even less in some locations. Furthermore the interaction of a spatially-variable precipitation intensity field with the spatially-distributed state of near-surface conditions determines the partitioning coefficient R. Other factors such as topography, vegetation cover and root systems all affect this coefficient. Clearly no comprehensive model of all such processes and interactions is feasible at the scale of the GCM grid. Early GCMs and some remaining operational GCMs opted to parameterize the bulk dependence of the ratio R on the grid-average surface soil moisture content. The model takes the form of linear or broken linear functions of the relative soil saturation s, $0 < s < 1$. For example in the NASA/ Goddard Institute for Space Studies (GISS) GCM,

$$R = \tfrac{1}{2} s \qquad (1)$$

(Hansen et al., 1983). In the NOAA/Geophysical Fluid Dynamics Laboratory (GFDL) GCM, R is taken as zero when the soil is not fully saturated and as unity when it is saturated. Similar linear or broken-linear functions have been defined for other GCMs; their parameters and coefficients are modified until they yield an adequate moisture and energy balance. Clearly these are models that attempt to represent the bulk response of the system without consideration of soil hydraulic properties and physical runoff generation mechanisms.

The evaporation from bare-soil and vegetated surfaces is treated in a similar fashion. The actual evaporation rate is taken as a fraction β of the energy-limited (potential) rate, $E = \beta E_p$. The efficiency β is also modelled as linear or broken-linear functions of soil saturation. In the GISS model, β is taken to be linearly related to soil saturation as in

$$\beta = s \qquad (2)$$

(Hansen et al., 1983). In the GFDL model, β is unity above a

critical soil saturation ($s = \tfrac{3}{4}$) and it is linearly related to s below this level.

Recently more sophisticated models of land surface hydrology have been introduced that focus attention on the presence of vegetation. The Biosphere-Atmosphere Transfer Scheme (BATS) and the Simple Biosphere Model (SiB) have been implemented in some GCMs (Dickinson et al., 1981; Sellers et al., 1986). These models represent the transfer of heat and moisture between various layers in the soil, canopy and atmosphere system by the potential flow through a series of resistances. These models are capable of capturing important aspects of the canopy control of the transpiration rate and furthermore they have developed increased sophistication in including the effects of soil texture on hydraulic flow.

In the parameterizations discussed thus far, the soil and vegetation surface over the GCM grid has been assumed to respond as a single hydrological unit; the forcing, the state of the surface and the response have all been taken to be uniform over the area constituting the GCM grid (of the order of tens to hundreds of thousands of square-kilometres). However, it is known that the hydrological partitionings under consideration result from the interaction between several spatially-variable fields (e.g. rainfall intensity, soil saturation, soil texture, slopes, etc.). The surface runoff results from both infiltration-excess and partial area contribution (rainfall over impermeable and saturated surfaces) mechanisms. The area-averaged evaporation rate is a combination of evaporation over subgrid areas that are either limited by energy or by moisture availability. The response of the large land area to atmospheric forcing must take into account these important physical factors.

In response to the need to include the effects of subgrid-scale spatial variability along with physically-based rather than empirical relations for surface hydrological processes, Warrilow et al. (1986) and Entekhabi and Eagleson (1989) used the statistical-dynamical approach to parameterizing the runoff process. Entekhabi and Eagleson (1989) furthermore extended the approach to include the bare-soil evaporation and transpiration processes.

The runoff at any location at the surface is generated by either infiltration-excess or partial area contribution. In the former case, runoff is the excess of the point rainfall intensity P over the local infiltration rate f^*. The infiltration rate is governed by the soil hydraulic properties which for unsaturated soils depend on both soil saturation s and soil texture. At any point within a grid or large land area one may write,

$$\text{Point surface runoff} = \text{Infiltration excess } (P - f^*) \qquad (3)$$

for $P > f^*$ and $s < 1$ and partial area (P for $s \geq 1$).

The point values of P and s over a large area may be regarded as belonging to a probability distribution function whose mean values $E[P]$ and $E[s]$ are the GCM grid values of

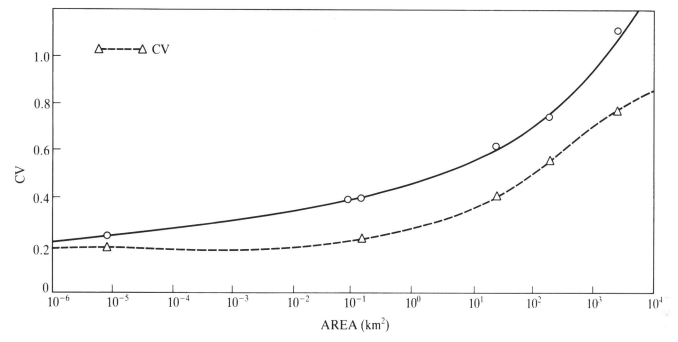

Fig. 1 Observed dependence of the coefficient of variation (*cv*) of near-surface soil moisture as a function of area (after Wetzel and Chang, 1988).

these variables. Thus the moist convection or large-scale supersaturation schemes yield $E[P]$ and water balance is updated by modifying $E[s]$ only. Entekhabi and Eagleson (1989) demonstrate that the probability density function for point precipitation may be shown to be exponential over a fraction κ of the GCM grid. Furthermore the soil saturation at the surface is gamma-distributed with mean $E[s]$ and a coefficient of variation parameter $\alpha = cv^{-2}$. Given the distribution functions for P and s, Entekhabi and Eagleson (1989) derived an analytical expression for the grid-mean runoff ratio to be used in water balance:

$$ R = 1 - \frac{\gamma\left[\alpha, \dfrac{\alpha}{E[s]}\right]}{\Gamma(\alpha)} + \frac{e^{-\kappa I(1-v)}\gamma\left[\alpha, \kappa I v + \dfrac{\alpha}{E[s]}\right]}{\left[\dfrac{\kappa I v E[s]}{\alpha} + 1\right]^{\alpha}\Gamma(\alpha)} \qquad (4) $$

where $\gamma(\ )$ and $\Gamma(\ )$ are the incomplete and complete γ functions, v is a given soil hydraulic parameter resulting from a physically-based infiltration function. The value of cv is dependent on the scale of the GCM grid and its topography. Owe et al. (1982) give estimates in a plot of cv versus logarithm of area (Fig. 1). Since this is a dimensionless statistic, it may also be estimated, without geophysical calibration, from remotely-sensed microwave irradiances that have known functional dependence on soil moisture. There is ongoing research on estimating values of cv based on observations. The dimensionless parameter I in eq. (4) is

$$ I = \frac{K_{sat}}{E[P]} \qquad (5) $$

where K_{sat} is the soil hydraulic conductivity at saturation and $E[P]$ is the grid mean precipitation intensity generated by GCM atmospheric computations.

The behaviour of R as a function of $E[s]$ and $E[P]$ is plotted in Fig. 2 for the case of a light textured soil ($K_{sat} = 7 \times 10^{-3}$ [m hr^{-1}] with $cv = 1$ and $\kappa = 0.6$ and 0.1 for a typical range of $E[P]$).

The GISS runoff ratio is independent of precipitation intensity (eq. (1)). The Entekhabi and Eagleson (1989) runoff ratio that includes subgrid-scale spatial variability, however, is sensitive to the precipitation regime. Furthermore, when the precipitation is spatially concentrated over the grid area (for example, moist convection events with limited spatial extent and intense short-period rain rate), the fraction of the total precipitation that cannot infiltrate increases and is therefore lost to runoff.

The fractional wetting parameter κ (representing the fraction of the grid square actually wetted by a precipitation event) exerts a strong influence in the parameterization of runoff. Further research has resulted in a robust measure of κ based on rain gauge observations. Gong et al. (1994) estimated κ over the continental US for each of the twelve months of the year.

Entekhabi and Eagleson (1989) also formulate an advanced parameterization of evaporation based on sub-

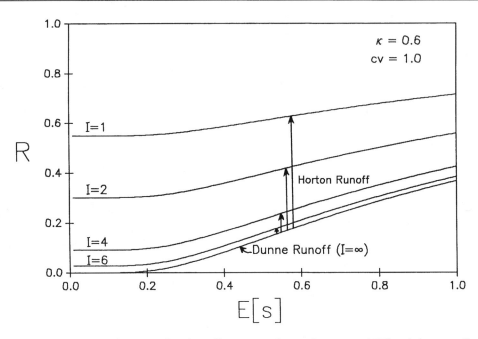

Fig. 2 The runoff ratio R for a large land area as a function of its mean surface moisture state ($E[s]$) and the normalized rainfall intensity I. The fractional coverage of the rainstorm is κ, and cv represents the coefficient of variation for the subgrid values of soil moisture.

grid-scale spatial variability and a physically-based exfiltration function. The parameter β in eq. (2) is separated into two components: β_s for bare soil evaporation and β_v for transpiration over vegetated areas. In both cases, the contributions of the grid fraction with the energy-limited evaporation regime and the grid fraction with water-limited evaporation regime are included. A physically-based desorption function is used in the β_s case; a parsimonious root soil-water extraction function is used in the β_v case. Using these definitions, the expected value of area evaporation or the grid evaporation rate is:

$$\beta = \frac{\Omega\gamma\left[\dfrac{1}{2m}+2+\alpha,\alpha\varepsilon^{-1}\right]-\dfrac{1}{2}\gamma\left[\dfrac{2}{m}+3+\alpha,\alpha\varepsilon^{-1}\right]}{\left[\Omega(\alpha\varepsilon^{-1})^{\frac{1}{2m}+2}-\dfrac{1}{2}(\alpha\varepsilon^{-1})^{\frac{2}{m}+3}\right]\Gamma(\alpha)}$$
$$+1-\frac{\gamma(\alpha,\alpha\varepsilon^{-1})}{\Gamma(\alpha)} \qquad (6)$$

Here ε is a dimensionless variable composed of soil saturation and hydraulic properties. The parameters m and Ω are defined as combinations of soil texture properties. Similarly, the transpiration efficiency function is derived as

$$\beta_v =$$
$$1+\frac{\gamma(\alpha+1,\alpha\varepsilon^{-1})-\alpha\varepsilon^{-1}\gamma(\alpha,\alpha\varepsilon^{-1})-\gamma(\alpha+1,\alpha W^{-1}+\alpha W^{-1}\gamma(\alpha,\alpha W^{-1})}{\Gamma(\alpha)(\alpha\varepsilon^{-1}-\alpha W^{-1})}$$
$$(7)$$

where W is a vegetation-type parameter. Entekhabi and

Eagleson (1989) may be referred to for the detailed definitions of the terms for all three R, β_s and β_v expressions.

Fig. 3 shows plots of β_s for light textured soil with $cv = 1$ over a typical range of potential evaporation. Similar expressions are derived for β_v (Entekhabi and Eagleson, 1989). The inclusion of subgrid-scale spatial variability and the incorporation of physically-based equations in the Entekhabi and Eagleson (1989) parameterization results in nonlinear dependencies between R, β_s and β_v on soil saturation, soil properties and the grid-mean precipitation forcing. It must be noted that the water balance is achieved by applying these expressions to the grid-mean forcing. This approach to incorporating spatial variability and physics-based equations into GCM landsurface parameterizations is considerably more efficient (in terms of computational costs) and significantly more parsimonious (in terms of the number of parameters) than subdividing GCM grids into smaller units, each with an independent hydrologic balance.

3. SIMULATIONS WITH THE GISS-GCM EQUIPPED WITH A NEW HYDROLOGICAL PARAMETERIZATION

Johnson et al. (1993) performed numerical experiments using the 8 × 10 degree GISS-GCM with seasonally fixed climate sea-surface temperatures. The seasonally fixed ocean temperatures are desirable when performing land-surface hydro-

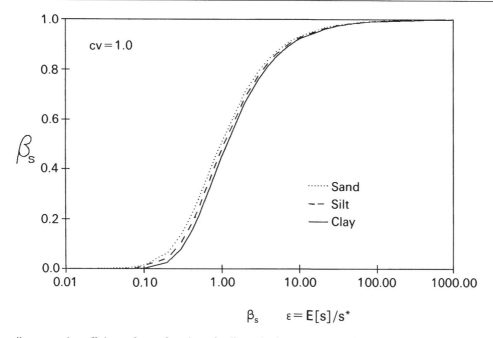

Fig. 3 The bare soil evaporation efficiency β_s as a function of a dimensionless parameter ε incorporating areal mean soil moisture and soil texture. Similar expressions are derived for the transpiration efficiency β_v. The symbol s^* denotes transitional relative soil saturation. If $s > s^*$ evaporation is climate controlled, if $s < s^*$ evaporation is soil controlled (after Entekhabi and Eagleson, 1989).

logical sensitivity tests. Low frequency transients introduced by a coupled ocean model may complicate the validation of the model climate. Run G-0 is the control case with the GISS-II GCM (Hansen et al. 1983) soil hydrology (an example is given by eq. (1)). Run G-3 included the Entekhabi and Eagleson (1989) parameterization as well as the Abramapolous et al. (1988) model for the diffusion of moisture between several soil layers.

Fig. 4 illustrates the comparison between the GISS-GCM with the empirical hydrologic parameterization (Run G-0), the GISS-GCM with the new hydrologic model (Run G-3) and available estimates based on observations. Run G-0 has grossly underestimated runoff and overestimated evaporation over the global landsurface (Fig. 4). The range of values of global land-surface runoff based on the available observations is from 27 to 35 cm annually. The GISS value does not fall in this range. Major improvements occurred in Run G-3 where runoff and evaporation fall within their observed ranges.

Figs. 5 and 6 show the zonally averaged values of evaporation and surface runoff over the land-surface for both the G-0 (Control) and G-3 simulations. Only land grids are included in order to avoid the masking induced by the large ocean fraction at most latitudes; as a consequence, at around 60° South where there is a small land fraction, the estimates become noisy. The zonally-averaged land-surface figures indicate that the model climate is improved for the GISS-GCM equipped with the Entekhabi and Eagleson (1989) parameterization. The runoff and evaporation rates resulting

from using the empirical formula (Run G-0) are inadequate when compared to estimates based on observations.

The model equipped with the new hydrological parameterization has improved the hydrological balance and furthermore it simultaneously achieves a more realistic heat balance as well. Johnson et al. (1993) compared the model results from an incremental implementation of the parameterizations. Furthermore they extended the comparison with observations to smaller regional scales as well. The reader is referred to Johnson et al. (1993) for details of these validation experiments.

4. CONCLUDING REMARKS

The proper simulation of water and energy partitioning at the land-surface has a strong influence on the model climate. Much of the forcing of atmospheric motions is due to the differential heating and heating gradients. The land-surface is responsible for the conversion of much of the solar energy to heat available for the atmosphere. The thermal and moisture capacities of the soil column and vegetation canopy result in the dampening of higher frequency variabilities in the atmosphere and they strongly influence the amplitude of the seasonal and diurnal cycles in the model climate.

The inclusion of hydrological parameterizations that partition the atmospheric forcing between storage and loss is thus of critical concern for atmospheric modelling. The realism with which subgrid-scale spatial variability and

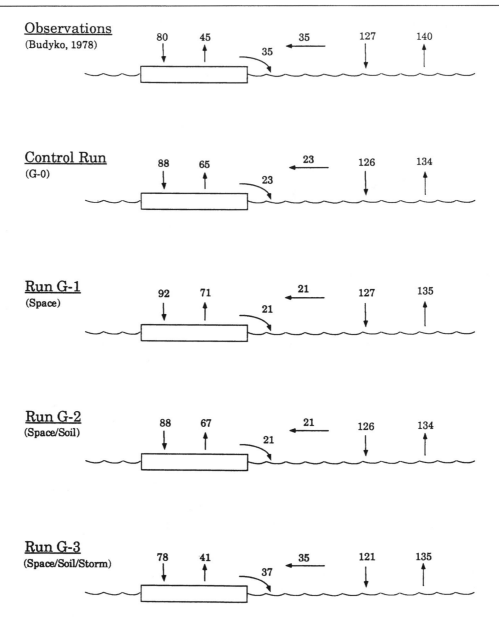

Fig. 4 Global hydrological balance (in cm/year) for simulations G-0 (Control) and G-3 (including improved hydrological parameterization). Comparison with Budyko (1978) estimates based on observations. Run G-0: GISS Model II as described in Hansen et al. (1983) using 8 × 10 degree grid resolution and fixed ocean temperatures. Run G-1: Same as G-0, except for new formulation of runoff coefficient R, bare soil, evaporation efficiencies, B_s, and transpiration efficiency, β_v (after Entekhabi and Eagleson, 1989). Fractional wetting parameter κ set equal to 1.0 for large-scale supersaturation rainfall and 0.6 for moist-convective rainfall. Run G-2: Same as G-1, except new soil moisture diffusion scheme of Abramopoulos et al. (1988) is used with transpiration from lower layers. No instantaneous upward diffusion or prescribed growing season as in GISS II. Run G-3: Same as G-2, except fractional wetting parameter κ for moist-convective rainfall set to 0.15.

physically-based relations of hydrology are used partially determines the realism of the hydrological parameterization. This paper has outlined the approach of Entekhabi and Eagleson (1989) to a computationally-efficient and parsimonious model. The efforts of Johnson et al. (1993) in providing observational evidence for improved water balance with the new hydrological parameterization are also outlined.

It is imperative that:

– analogous improved models be developed for snow and ice processes,
– schemes be developed to estimate the model parameters using remotely-sensed and surface observations, and
– serious efforts be made to compare the model global and regional heat and water balance with improved estimates based on observations.

Zonally Averaged Evaporation
(Landsurface Only)

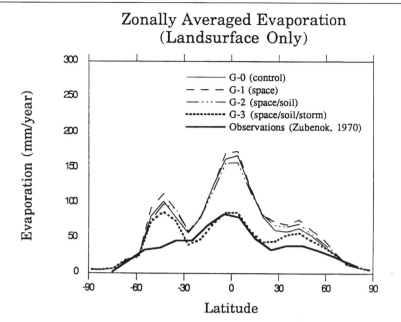

Fig. 5 Zonally averaged annual evaporation over land grids. Observations from Zubenok (1970) estimates.

Zonally Averaged Surface Runoff

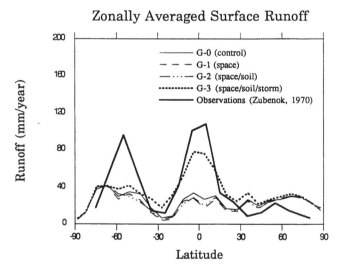

Fig. 6 Zonally averaged annual surface runoff over land grids. Observations from Zubenok (1970) estimates.

REFERENCES

ABRAMOPOULOS, F., C. ROSENZWEIG, and B. CHOUDHURY, 1988: Improved ground hydrology calculations for GCMs – Soil water movement and evapotranspiration, Journal of Climate, 19, 921–941.

BUDYKO, M.I., 1978: The heat balance of the earth, in Climatic Change, edited by J. Briggin, Cambridge University Press, Cambridge, pp. 85–113.

DICKINSON, R.E., J. JAGER, W.M. WASHINGTON, and R. WOLSKI, 1981: Boundary Subroutine for NCAR Global Climate Model, NCAR Technical Note 173TIA.

ENTEKHABI, D. and P.S. EAGLESON, 1989: Landsurface hydrology parameterization for atmospheric general circulation models including subgrid scale spatial variability, Journal of Climate, 2(8), 816–831.

GONG, G., D. ENTEKHABI and G.D. SALVUCCI, 1994: Regional and seasonal estimates of fractional storm coverage based on station precipitation observations, Journal of Climate, 7(10), 1495–1505.

HANSEN, J., G. RUSSELL, D. RIND, P. STONE, A. LACIS, S. LEBEDEFF, R. RUEDY and L. TRAVIS, 1983: Efficient three-dimensional global models for climate studies: models I and II, Monthly Weather Review, 111(4), 609–662.

JOHNSON, K.D., D. ENTEKHABI and P.S. EAGLESON, 1993: The implementation and validation of improved landsurface hydrology in an atmospheric general circulation model, Journal of Climate, 6(6), 1009–1026.

OWE, M., E.B. JONES and T.J. SCHMUGGE, 1982: Soil moisture variation patterns observed in Hand County, South Dakota, Water Resources Bulletin, 18(6), 949–954.

PHILIP, J. R., 1957: The theory of infiltration: 1. The infiltration equation and its solution, Soil Science, 83, 345–357.

SELLERS, P.J., Y. MINTZ, Y.C. SUD, and A. DELCHER, 1986: A simple biosphere model (SiB) for use within General Circulation Models, Journal of Atmospheric Science, 43(6), 505–531.

WARRILOW, D.A., A.B. SANGSTER, and A. SLINGO, 1986: Modelling of land surface processes and their influence on European climate, UK Meteorological Office, DCTN 38, 94 pages, available from the UK Meteorological Office.

WETZEL, P.J. and J.-T. CHANG, 1988: Evapotranspiration from nonuniform surfaces: A first approach for short-term numerical prediction, Monthly Weather Review, 116, 600–621.

ZUBENOK, L.I., 1970: Improved scheme of the water balance of continents in heat balance, edited by T.G. Berlyand, Hydrometeorological Press, Leningrad.

Modelling the hydrological response to large scale land use change

A. HENDERSON-SELLERS[1], K.McGUFFIE[2] and T.B. DURBIDGE[1]

[1] *Climatic Impacts Centre, Macquarie University,*

North Ryde, NSW 2109,

Australia

[2] *Department of Applied Physics, University of Technology,*

Sydney, PO Box 123,

Broadway, NSW 2007,

Australia

ABSTRACT Macroscale hydrological modelling is currently conducted using Global Climate Models (GCMs) coupled to a range of Soil-Vegetation-Atmosphere Transfer Schemes (SVATs). The most extreme type of simulation involves massive land use change. This paper reports on the results of a tropical deforestation experiment in which the tropical moist forest throughout the Amazon Basin and SE Asia has been replaced by a scrub grassland in a version of the NCAR Community Climate Model (Version 1) which also incorporates a mixed layer ocean and the Biosphere-Atmosphere Transfer Scheme (BATS). In the Amazon we find a smaller temperature increase than did all other previous experiments except Henderson-Sellers and Gornitz (1984); indeed temperatures decrease in some months. On the other hand, we find larger hydrological responses than all earlier experiments including runoff decreases and a larger difference between the changes in evaporation and precipitation which indicate a basin-wide decrease in moisture convergence. Disturbances extend beyond the region of land-surface change causing temperature reductions and precipitation increases to the south of the deforested area in S America. Changes to the surface climate in the deforested area take between 1 and 2 years to become fully established although the root zone soil moisture is still decreasing at the end of a 6-year integration. Besides temperature and precipitation, other fields show statistically significant alterations, especially evaporation and net surface radiation (both decreased). An important question raised by this type of simulation concerns the appropriateness of the microhydrological process models employed in SVATs to the GCMs in which they are currently used.

1. SIMULATING THE EFFECTS OF LARGE SCALE LAND USE CHANGE

Global climate (GCM) modellers could be seen as charlatans: taking theory from the microscale and applying it (unashamedly) to their macroscale models (Fig. 1 and Becker, 1993, this volume). In doing this, they seem to have skipped over the meso (or basin-) scale which is of paramount importance to hydrologists and which contains the exciting dynamics of the planetary boundary layer (PBL) and of heterogeneous surface-atmosphere exchanges (Avissar, 1993, this volume). An important question is whether this application of local-scale observations and theory to the large areas of GCM grids is valid. This paper is based upon the assumption that such models are both valid *and useful* at least when they are applied to *large* areas of land use change and to the *regional to global scale* climatic and hydrological impacts.

The claim has been advanced that the removal of tropical rainforests will substantially alter the climate: by adding CO_2

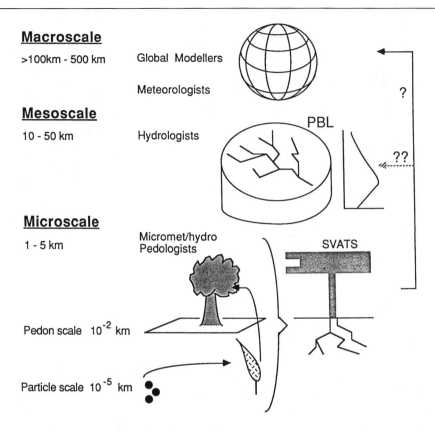

Fig. 1 Global climate modellers freely utilize microhydrological and micrometeorological concepts and parameterizations in their coarse resolution models. This exploitation is based on the assumption that scaling of hydrological and other properties is reasonable.

to the atmosphere, thereby enhancing the greenhouse effect, by increasing the global surface albedo or by modifying other properties, such as the evaporation (Bolin, 1977; Woodwell et al., 1978; Sagan et al., 1979; Hampicke, 1980; Potter et al., 1981; Shukla and Mintz, 1982). Indeed the 'devastating' impact on hydrology is often cited as an important reason for protecting tropical forest. For example, the Australian Rainforest Memorandum (1991, p8) states: 'Rainforests generate local and global rainfall, receiving 50% of all the rain which falls on land. Nobody knows how much primary or secondary rainforest must be preserved to ensure the continuity of their characteristic climatic cycles. Current changes in the weather patterns of the Ivory Coast and the once forested coastal areas of south-east Brazil suggest that tropical deforestation is leading towards climatic disturbances of disastrous magnitude'.

Besides these questions of environmental concern, the study of tropical deforestation offers an excellent route to improving treatments of land/climate interaction in climate models especially aspects of the surface hydrology. The GCM studies published to date are Henderson-Sellers and Gornitz (1984), Dickinson and Henderson-Sellers (1988), Lean and Warrilow (1989) and Shukla et al. (1990)/Nobre et al. (1991).

1.1 Formulation of hypotheses and experimental design

The four published GCM simulations of the impacts of tropical deforestation differ in almost every respect from one another: model type, resolution, time of simulation, prescribed deforestation and, naturally, results. These experiments and their results are summarized in the first four columns of Table 1. Only the first (Henderson-Sellers and Gornitz, 1984) could be seen as having been temporally extensive enough to permit a complete assessment of the climatic response, although being the coarsest in resolution and using a simpler land-surface scheme than employed in the other three simulations.

All four GCM studies have focussed on the impacts of deforestation in the Amazon region of South America, presumably because this is the largest remaining area of tropical moist forest, although Henderson-Sellers and Gornitz (1984) undertook a preliminary estimate of the possible impact of global deforestation and Shukla et al. (1990) comment on the possible impact of African deforestation. Despite the acknowledged devastation of deforestation in much of the South East Asian region there have been no previous experiments using GCMs to investigate possible impacts of tropical deforestation in SE Asia or possible links to impacts caused by Amazonian deforestation.

Table 1. *Comparison of previous GCM simulations with this study.*

Reference*	H-S & G (1984)	D & H-S (1988)	Lean & Warrilow (1989)	Shukla et al.[†] (1990)	This exp (1992)
Model:					
GCM:	GISS grid	CCM0B spectral (R15)	UKMO grid	NMC spectral (R40)	CCM1-Oz grid (R15)
Resolution:	$8° \times 19°$	$4.5° \times 7.5°$	$2.5° \times 3.75°$	$1.8° \times 2.8°$	$4.5° \times 7.5°$
Surface:	2-layer hydrol.	BATS	canopy interception	SiB	BATS
Clouds:	55%	30% (conv.)	(not given)	fixed zonal means	Slingo (1989)
Ocean:	fixed transports	fixed SSTs	fixed SSTs	perpetual December	mixed layer model
Length: control/def.	20/10 yr	3/1 yr	3/3 yr	1/1 yr	6/6 yr
Results:					
ΔT (°C)	0	+3	+2.4	+2	+0.6
ΔP (mm)	−220	0	−490	−640	−588
ΔE (mm)	−164	−200	−310	−500	−232
Δ moisture flux convergence	little change	not known	decreased	decreased	decreased esp. in January

Identified problems:

H-S & G: poor spatial resolution: simple surface; no canopy

D & H-S: short simulation; too large net surface radation; too large canopy water store

Lean & Warrilow: evaporation too large in control; 71% evaporation from canopy in control cf. 25% observed

Shukla et al. : short simulation; error in ocean temps; no cloud feedback

* Henderson-Sellers and Gornitz (1984) is H-S & G (1984)
  Dickinson and Henderson-Sellers (1988) is D & H-S
† reported in more detail in Nobre et al. (1991)

Despite the differences amongst past models in formulations and detailed results, these GCM experiments indicate: (i) that changes occur in the continental surface climate over the region of deforestation but that the magnitude (and even the direction) of these changes remains uncertain; (ii) that deforestation may cause atmospheric circulation changes which may modify vertical motion over the deforested region and hence advection to/from nearby regions; (iii) that, as a consequence of (ii), there exists the possibility that regions adjacent to the deforested area could exhibit climatic disturbances; and, finally, (iv) modified vertical motion and/or atmospheric convergence patterns might signal a disturbance to aspects of the general circulation especially the Walker and Hadley cells which derive their energy from warm, moist air ascending in equatorial regions. The confidence with which it is sensible to anticipate these consequences of tropical deforestation decreases from (i) to (iv) since all previous studies indicate local surface changes and no previous simulations have reported positive identification of disturbances to the large scale circulation of the Earth.

Our GCM simulations have been designed to bridge some of the gaps between these earlier simulations and to investigate all four hypothesized climatic impacts. The same spatial resolution as used by Dickinson and Henderson-Sellers

(1988) (spectral R15) and an upgraded version of the same GCM (NCAR CCM) is used but, here, also including a slab ocean. The integrations are for a period of six years, intended not only to be long enough to overcome the effects of initial conditions and instantaneous forest removal on the land-surface climate but also to permit multi-year ensemble averages to be constructed and evaluated. Results are analysed first for the deforested region, then for disturbances external to one deforested region and finally to try to assess the possibility of teleconnections.

1.2 Henderson-Sellers and Gornitz (1984)

This was the first numerical deforestation experiment with a GCM. Simpler models suggested that global responses to albedo change may have been over-estimated by Sagan et al. (1979). Amazonian deforestation was modelled by changing albedo, roughness length and the soil moisture holding capacity. The prescribed changes were intended to capture the characteristics of a change in vegetation cover from tropical forest to grassland. The model used was the Goddard Institute for Space Studies (GISS) finite difference model at 8° latitude by 10° longitude resolution. They undertook a 10 year simulation and used the last 5 years' results to compare with the last 5 years of a 20 year control run.

Henderson-Sellers and Gornitz (1984) used a low resolution model (8 × 10 degrees) to increase the length of the simulations and the realism of oceanic parameters. Their ocean sub-model included ocean temperature calculation, fixed ocean transports and simulation of sea ice to allow investigation of global teleconnections. A total area of 4.94×10^6 km^2 was deforested, equal to about 35–50 years worth of global tropical deforestation, concentrated in the Amazon basin. The focus of the simulation was large scale and local changes in all parameters plus global teleconnections. This simulation remains the longest study (in terms of simulated years) and the most complete in the sense that a type of mixed layer ocean model was coupled to the atmospheric model.

On the other hand, the land surface scheme employed was the simplest used to date, as it included only a 2-layer soil moisture model, coupled instantaneously to represent the role of vegetation unless the upper layer is wetter than the lower, in which case the layers are coupled on a one-day time scale (analysed in Dickinson and Henderson-Sellers, 1988). The GCM included an interactive cloud scheme; and differentiation of convective and large scale cloud cover.

The analysed parameters included temperature, rainfall, planetary albedo, evaporation, cloud amount and available soil moisture. No significant large scale or global scale effects were observed. Over the basin, there was no surface temperature change; decreased rainfall (0.6 mm d^{-1}); decreased evaporation (0.4–0.5 mm d^{-1}); decreased cloud cover (5–15%); decreased soil moisture availability; increased planetary albedo (1–1.5%).

Henderson-Sellers and Gornitz (1984) concluded that the effects of continental hydrology must be included to give realistic simulations of the tropical environment and especially of tropical deforestation. They found that, while immediate large scale effects were significant, no teleconnections or global scale effects could be discerned (contrary to earlier suggestions, e.g. Sagan et al., 1979). More recently, this conclusion has been challenged (e.g. Shukla, 1987 and Shukla et al., 1990).

This first GCM simulation suffered from the problems commonly associated with the GISS climate model at the time: low resolution; failure to represent subgrid scale processes adequately (e.g. clouds, topography); poorly understood feedbacks (e.g. tropical ocean temperatures); poor prediction of precipitation in the control run, considerably lower than observed. The land-surface scheme did not include a canopy, which is significant for the surface energy budget, especially the diurnal cycle of surface fluxes (Dickinson and Henderson-Sellers, 1988). At the time of the experiment (1981/82) there had been little attempt to apply

knowledge about the actual microphysical processes (especially their magnitudes) in tropical rainforest canopies (e.g. Shuttleworth, 1988; Dickinson, 1989) to GCM simulations.

1.3 Dickinson and Henderson-Sellers (1988)

This study attempted to clarify the role of the tropical forest canopy in controlling the surface energy balance. It also briefly reviews the then state of land-surface schemes for use in GCMs. Dickinson and Henderson-Sellers (1988) performed stand-alone tests of the Henderson-Sellers and Gornitz (1984) 2-level soil moisture scheme (a 2-level bucket) and compared the results with those from the Biosphere-Atmosphere Transfer Scheme (BATS) of Dickinson et al. (1986) under tropical forest conditions. BATS gave more realistic results for sensible and latent heat fluxes and, generally, a 'more responsive and diurnally correct' representation of energy and moisture fluxes.

Dickinson and Henderson-Sellers (1988) modelled tropical deforestation in the Amazon by changing maximum fractional vegetation cover, difference between maximum fractional vegetation cover and cover at 269 K, roughness length, depth of total soil layer, rooting ratio (density of vegetation roots), vegetation albedo at wavelengths less/more than 0.7 micrometres, minimum leaf area index, leaf dimension, and light sensitivity factor to represent the change from tropical moist forest to impoverished grassland (other BATS parameters were the same for forested/deforested conditions). In addition, the soil texture and colour were made finer and lighter to represent deforested conditions better. The parameters used are listed in the first two columns of Table 2.

They used the NCAR CCM0B (spectral) at R15 resolution (~4.5° latitude by 7.5° longitude), undertaking a 13 month integration, initialized from the second year of a control simulation. The model had not been allowed to reach equilibrium first. The CCM0B included cloud formation as a function of relative humidity and differentiated between convective and large scale cloud cover fractions.

The numerical experiment comprised altering the specified land-surface characteristics of 18, 4.5° latitude × 7.5° longitude grid squares in Amazonia from evergreen broadleaf to impoverished grassland, with finer and lighter coloured soils. Dickinson and Henderson-Sellers (1988) note that the CCM0B's seasonal pattern of precipitation was fairly satisfactory, but that modelled precipitation was greater than observed in some parts of the study area. The BATS land-surface scheme was more realistic than that used in the earlier experiment since it includes precipitation interception, evaporative loss, leaf drip, moisture uptake by plant roots and stomatal resistance.

Table 2. *Values for the forest (Forest) and deforested (Def.) experiments for the 24 ecotype and soil factors in BATS as compared to those employed by Dickinson and Henderson-Sellers (D & H-S) (1988).*

Factor	D & H-S (1988)		This project	
	Forest	Def.	Forest	Def.
A maximum vegetation cover	0.9	0.8	0.9	0.8
B difference between maximum vegetation cover and cover at 269 K	0.5	0.3	0.5	0.3
**C vegetation roughness length (m)	2.0	0.05[+]	2.0	0.2
D total soil depth (mm)	—	—	1.0E+4	1.0E+4
E upper soil depth (mm)	100.0	100.0	100.0	100.0
F rooting ratio	12	10[x]	0.8	0.5
**G vegetation albedo < 0.7 μm	0.04	0.08	0.04	0.08
*H vegetation albedo ≥ 0.7 μm	0.2	0.3	0.2	0.3
I minimum stomatal resistance (s m^{-1})	250.[@]	250.[@]	150.[@]	200.[@]
*J maximum leaf area index	6.0	6.0	6.0	6.0
K minimum leaf area index	5.0	0.5	5.0	0.5
*L stem area index	2.0	2.0	2.0	2.0
M inverse square root of leaf dimension (m$^{-\frac{1}{2}}$)	5.0	5.0[!]	5.0	5.0
*N light sensitivity factor (m^2 W^{-1})	0.03	0.01[=]	0.06	0.02
O soil porosity§	0.6	0.66	0.6	0.66
P minimum soil suction (mm)§	200	200	200	200
Q saturated soil hydrologic conductivity (mm s^{-1})§	1.6E-3	0.8E-3	1.6E-3	0.8E-3
R ratio of saturated soil thermal conductivity to that of loam§	0.8	0.7	0.8	0.7
S soil diffusivity exponent§	9.2	10.8	9.2	10.8
T soil moisture content relative to saturation at which transpiration ceases§	0.487	0.542	0.487	0.542
U soil colour class§	4	2	4	2
V vegetation displacement height (m)	—	—[†]	18.0	0.0
W depth of root zone (mm)	1500.0	1000.0[†]	1500.0	1000.0
X exponent in runoff term	4.0	2.0[$]	4.0	4.0

Notes:

[†] Soil depths and displacement height dealt with differently after 1988. New total soil depth (equal to 10 m for all ecotypes) introduced and depths previously listed as 'total soil' now found in 'root zone'. New displacement height introduced but roughness lengths remain as before.

* Most important effects as identified in a suite of factorial experiments (Henderson-Sellers, 1992).

** (As for *) individually important and also an important two-factor interaction.

[+] Dickinson changed this value to 0.3 after 1988.

[@] Dickinson modified these values to 150 (forest) and 200 (deforested) after 1988.

[=] Dickinson modified these values by a factor of 2 after 1988.

[!] H-S et al. 1988 give this value as 2.0.

[x] This ratio of roots in upper to total soil layers parameter form changed after 1988: a value of 1.0 implies all roots in upper layer, whereas 0.0 means none of roots in upper soil. New values are achieved by multiplying old values by param E/param W which gives a value of 1.0 for deforested vegetation type 19. The two largest values in the BATS code are 0.9 given for desert and tundra. The replacement vegetation type (19) is more like 'deciduous shrub' (type 17) which has a value of 0.5. NB The roots in question are only moisture tapping roots not those which deliver nutrients to the plants nor those used primarily to anchor the plant against wind.

[$] The runoff formulation was altered by Dickinson & Henderson-Sellers (1988): they used a value of 2 for the deforested environment. The standard BATS code (all other ecotypes except ice which takes a value of 1) uses a value of 4.

§ To simulate deforestation the soil texture class (params O to T) made finer by two categories (up to a max. of finest class = 12). Typically change is from 10 to 12 (clay). In addition, the soil colour is made lighter by two classes; typically from 4 to 2.

Dickinson and Henderson-Sellers (1988) included analysis of the following parameters: precipitation, surface air temperature, surface soil temperature, total soil temperature, canopy interception, evaporation, soil moisture and runoff. The only results reported were for the Amazon region itself although the published maps of climatic responses suggested that the region to the south of the deforested Amazon Basin exhibited disturbances following the forest removal. Impacts in the deforested area itself included a rainfall response which was noisy with no systematic large scale change; an increase in surface air, surface soil and total soil temperature of approximately 2 K; decreased canopy interception except in August; decreased evaporation except in September; overall increase in surface and total runoff; and slight decreases in soil moisture.

The authors concluded that changes in surface roughness interacting with the canopy hydrology are significant in determining the model response to deforestation and in particular are largely responsible for the changes in surface temperature. They also noted that the reduction of the soil's water-holding capacity might be important during droughts. In common with the other papers reviewed here, Dickinson and Henderson-Sellers (1988) also concluded that the inclusion of more realistic land-surface schemes is essential to the proper modelling of deforestation.

Dickinson and Henderson-Sellers' (1988) simulations have subsequently been reviewed by Dickinson (1989) and Shuttleworth and Dickinson (1989) who point out that their experiment may have exaggerated the effects of deforestation on evapotranspiration, especially during the wet season, due to overestimation of interception (Dickinson, 1989) and excessive solar radiation incident at the surface (Shuttleworth and Dickinson, 1989).

Dickinson and Henderson-Sellers (1988) themselves comment that too little was known at the time of the initiation of their simulation about tropical forest microprocesses to prescribe adequately the changes made to BATS and that the very short integration period (13 months) means that the results they report were in principle a transient response to the applied disturbance. They also raise the question whether the warmer surface and longer dry season might prevent regrowth of tropical forest following Amazonian deforestation.

1.4 Lean and Warrilow (1989)

This paper reported the response of the UK Meteorological Office (UKMO) GCM: a finite grid model with a spatial resolution of 2.5° latitude × 3.75° longitude. The simulations built on an earlier (unpublished) study by Wilson (1984), the major differences between the experiments being the improved land-surface scheme and the longer simulation periods for both control and deforestation experiment. In

addition, Lean and Warrilow (1989) note that the cloud prediction scheme in the UKMO model had been improved prior to their simulation and that their land-surface scheme included a statistical parameterization of subgridscale variability in precipitation. The land-surface scheme used employs a four-layer temperature calculation, a parameterization of canopy interception of precipitation and its re-evaporation; and calculation of surface and subsurface runoff.

The numerical experiment comprised altering the surface characteristics of the 1° × 1° land-surface points in the Wilson and Henderson-Sellers (1985) data set from tropical forest to savannah. The simulation period was 3 years for both the control and the deforestation experiment. In addition, Lean and Warrilow (1989) conducted a further two eight-month simulations with the UKMO GCM to try to establish the relative importance of prescribed albedo and roughness length changes.

Lean and Warrilow (1989) reported only regional scale climatic changes which include temperature increased by 2.4 K; precipitation decreased by 1.34 mm d^{-1}; evaporation decreased by 0.85 mm d^{-1}; surface runoff decreased; available soil moisture decreased by 58.7%. They also reported a decrease in the moisture flux convergence over the Amazon region and noted that, while the reduction in evaporation and the increase in temperature were primarily caused by the prescribed change in the surface roughness length, the decrease discovered in the moisture flux convergence was caused by the prescribed albedo increase.

Lean and Warrilow (1989) claimed that their simulations indicated a larger climatic impact resulting from Amazonian tropical deforestation than previous studies. Their experiment suffered from many of the handicaps described elsewhere including an exaggerated seasonality in precipitation in the Amazon and too large a percentage of the canopy-intercepted moisture being re-evaporated (cf. Shuttleworth and Dickinson, 1989).

1.5 Shukla, Nobre and Sellers (1990)

This simulation (reported in Nobre et al., 1991) is the most recent GCM study and the experiment conducted with the highest resolution GCM. The authors modelled deforestation by changing albedo, roughness length, stomatal resistance, root system characteristics and soil moisture capacity by modelling the morphological, physiological and physical parameters in the SiB land-surface scheme (Sellers et al., 1986). The model used was the NMC GCM (spectral) model at R40 (1.8° × 2.8°) resolution (Sela, 1980; Kinter et al., 1988). This model has: 18 levels; explicit diurnal cycle; prescribed large scale topography and SST fields fixed at December values; prescribed cloudiness from seasonal means, with convective/large scale cloud differentiation.

The simulation experiment comprised two one-year integrations for the control and deforestation realizations. The area deforested comprised the whole of the Amazon basin and the authors reviewed the local, large scale and global effects of the deforestation. The local climatic impacts include surface soil temperature increased 1–3 K; 26% decrease in precipitation; 6% decrease in precipitable water; 30% decrease in evapotranspiration; 18% increase in evapotranspiration minus precipitation; and decreased runoff.

Shukla et al. (1990) emphasize the decrease in the moisture convergence over the Amazon Basin. At the large scale they found some perturbations over North America but comment that these could not be ascribed to the deforestation. They found no significant global effects. Shukla et al. (1990) conclude that the reduction in precipitation over Amazonia which they find, and which is larger than the corresponding reduction in evapotranspiration, implies that the result of deforestation may be a longer dry season, perhaps making the deforestation self-perpetuating.

In addition to the problems common to all global climate model simulations, the short integration lengths mean that equilibrium had not been achieved after deforestation. The authors are surprisingly confident in the realism of the changes they detect in the Amazon Basin hydrology despite the fact that the model which they use employs zonally and seasonally prescribed cloudiness and constant (December values) sea-surface temperatures. Their conclusions should be compared with the detailed reviews of the difficulty of achieving adequate simulations of surface-incident solar radiation (e.g. Dickinson, 1989; Shuttleworth and Dickinson, 1989) and the *interactive* nature of the surface hydrology and the radiation/cloud parameterizations (cf. Lean and Warrilow, 1989; Henderson-Sellers and Gornitz, 1984).

2. SIMULATING THE CLIMATIC EFFECTS OF DEFORESTATION IN THE AMAZON BASIN AND IN SE ASIA

We use an updated version of the currently available NCAR Community Climate Model (CCM1). A full description of CCM1 is given in Williamson et al. (1987) and circulation statistics from seasonal and perpetual January and July simulations of the standard version of CCM1 are given in Williamson and Williamson (1987). CCM1-Oz is a modified version of CCM1 which includes the current version of the Biosphere-Atmosphere Transfer Scheme (BATS1E) and a mixed-layer, slab ocean of 50 m depth. The mixed-layer ocean model includes a three-layer ice model subcomponent and a standard q-flux scheme to correct for ocean advection of energy and the prescription of a fixed mixed-layer depth. CCM1-Oz includes a number of modifications to the physics subroutines including the Slingo clouds and radiation updates (Slingo, 1989). The model simulates full seasonal

and diurnal cycles. Review of a number of standard global fields shows that the general circulation of the atmosphere is well simulated.

The experiment conducted is to modify the specified land use in a total of twenty seven R15 grid elements. The locations of the areas where deforestation is simulated are represented graphically in Fig. 2. In all cases the ecotype was changed from tropical moist forest to scrub grassland, the soil texture made finer by two classes and the soil colour made lighter by two classes. For Amazonia, eighteen tropical forest grid locations were modified, while in SE Asia, nine points were modified.

The landscape after tropical deforestation is difficult to specify but previous experiments have used a degraded scrub-grassland (cf. Dickinson and Henderson-Sellers, 1988; Shukla et al., 1990). Here we follow a similar procedure. We imagine this ecotype to be a tall grass-covered scrubland with a few large trees and a partial shrub understorey, perhaps the resultant vegetation after major deforestation and possible attempts at agricultural 'development'. The soil has almost recovered from soil compaction. The vegetation cover is almost complete except for some areas where soil erosion forms scars which have not revegetated. The roughness length must be significantly reduced from the forest value of 2 m. Following Wieringa's (1993) data and the following assumptions we chose to reduce it to 0.2. The effective roughness length is dominated by the roughest element in a given area. Wieringa's (1993) data suggest that the distribution of shrubs and trees bias the choice of roughness. Wieringa (1993) suggests 0.3 for shrubs and bush land, but we assume that the shrubs are sparse but clumped and so we further reduce 0.3 to 0.2 m. This deforested roughness length is substantially larger than that of any of the earlier experiments.

There are 24 parameters used to describe the vegetation and soil characteristics for the BATS submodel at every continental surface point (Table 2). These parameters have been reviewed and a series of factorial experiments used to try to determine their relative importance in tropical climates prior to conducting these simulations (Henderson-Sellers, 1992). Many proposed changes are to offer some consistency with previous experiments and/or to adhere, at least in part, with the ecological sense of the variable as well as its code implications.

Analysis of the regional scale to large scale impacts of this instantaneous deforestation is undertaken in two ways: with reference to spatially averaged results from three regions which together encompass the deforested areas and with reference to maps of South America and SE Asia. Fig. 3 shows the location of the three regions used: Region 1 (7 grid points, 5 forest) comprises the northern Amazon Basin where the majority of the rainfall occurs in mid-year; Region 2 (9 grid points, all forest) comprises the southern Amazon

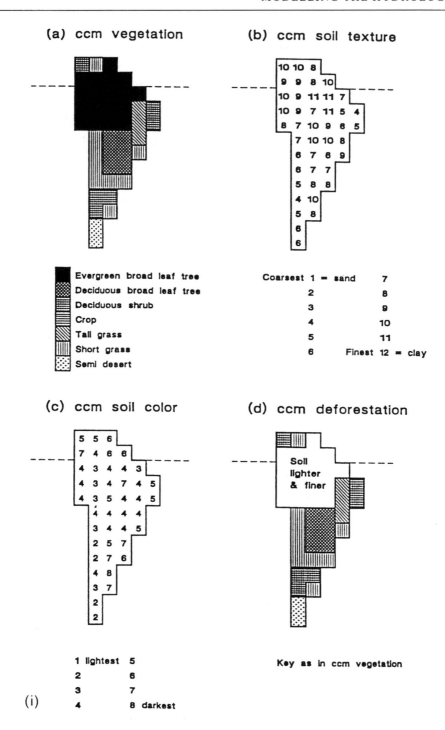

(a) ccm vegetation

(b) ccm soil texture

Evergreen broad leaf tree
Deciduous broad leaf tree
Deciduous shrub
Crop
Tall grass
Short grass
Semi desert

Coarsest 1 = sand   7
          2   8
          3   9
          4   10
          5   11
          6   Finest 12 = clay

(c) ccm soil color

(d) ccm deforestation

Soil
lighter
& finer

1 lightest   5
2   6
3   7
(i)   4   8 darkest

Key as in ccm vegetation

Fig. 2(i) Vegetation type (a), soil texture (b), soil colour (c) and the result of deforestation (d) for South America.

Basin in which the rainy season extends from October to March and Region 3 (8 grid points, 7 forest) comprises almost all the deforested area in SE Asia. The South American response was separated into two parts of the Amazon Basin because the precipitation seasonality is 6 months out of phase such that basin-wide averaging may suggest a better representation of the control climate than really exists. Although the same is true for SE Asia, the deforested region is smaller so it was decided to try to establish a response from the whole region in the first instance.

2.1 Regional-scale impacts in South America

Two regions are considered: the northern and southern Amazon Basin (see Fig. 3). Region 2 is the larger of these and consists solely of grid elements (9 in total) that have been deforested and is here considered first. Region 2 is also used to typify the response of the Amazon in Table 1. From this

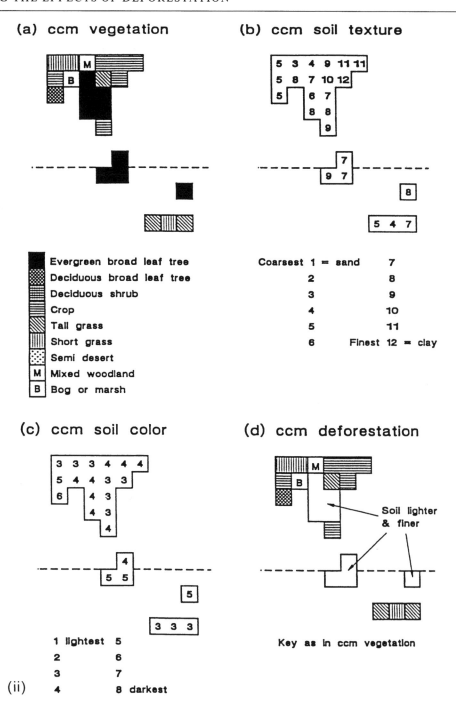

(a) ccm vegetation

(b) ccm soil texture

Coarsest 1 = sand
2
3
4
5
6   Finest 12 = clay

7
8
9
10
11

Evergreen broad leaf tree
Deciduous broad leaf tree
Deciduous shrub
Crop
Tall grass
Short grass
Semi desert
M Mixed woodland
B Bog or marsh

(c) ccm soil color

(d) ccm deforestation

Soil lighter & finer

Key as in ccm vegetation

1 lightest   5
2   6
3   7
(ii)   4   8 darkest

Fig. 2(ii) As Fig. 2(i) but for SE Asia.

table it can be seen that our annually-averaged results for Region 2 are in general agreement with those of previous studies. Two aspects of particular interest are the smaller temperature increase than all the previous simulations except Henderson-Sellers and Gornitz (1984) and a larger difference $\Delta(E-P)$ than in any previous experiment. The latter feature indicates that there is a basin-wide decrease in moisture convergence as a result of deforestation (since $E-P=$ moisture divergence). This result agrees with the findings of Shukla et al. (1990, also reported in Nobre et al., 1991) but

our response (of decreased convergence) is stronger than theirs and we find that the largest change in $(E-P)$ occurs in the rainy season (November–February). It must be noted that here we are using Region 2 as typical of the Basin as a whole although the decrease in convergence is found over the whole region as described below. Overall, these results differ most from those of the 'companion' GCM experiment conducted by Dickinson and Henderson-Sellers (1988) who found a large temperature increase and little or no precipitation change.

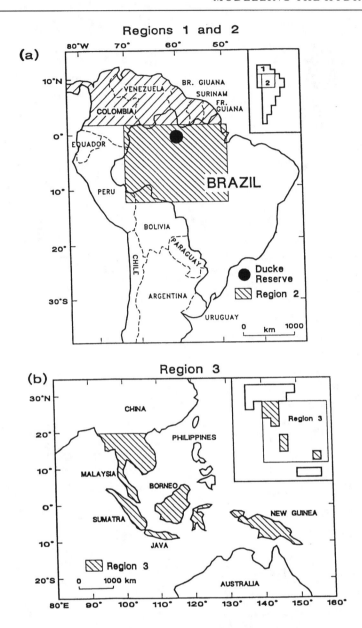

Fig. 3 Locations of the three regions used for analysis: Region 1 comprises 7 land grid elements in the northern Amazon Basin; Region 2 comprises 9 land grid elements in the southern Amazon Basin and Region 3 comprises 8 land grid elements in the peninsula and islands of SE Asia. The location of the Ducke Reserve is also shown (D), 2°57′ S, 59°57′ W.

Fig. 4 depicts the seasonal cycle of a selection of climatic parameters shown as five-year averages derived from both the control and deforestation experiments. The decrease in total precipitation is seen (Fig. 4(a)) to be significant (outside the 95% confidence limits) on the control in all months except May, July and September, that is, both the smaller decrease in the dry season and the larger decrease in the rainy season are significant because the year-to-year variability is smaller in the dry season. Rainfall is seen to decrease more markedly in the rainy season (November–April) (Fig. 4(a)). During this period all but one month has over 30% decrease in precipitation whereas between May and October only one month (August) has a decrease of this magnitude and two

months (May and September) show less than a 7% decrease. Similarly, the temperature decreases in the early part of the dry season (May–June) (Fig. 4(b)) are as significant as the increases in the rest of the year; evaporation decreases are significant all year (Fig. 4(c)) whereas sensible heat differences are only significant in January–April. The total runoff (Fig. 4(d)) and net radiation are also significant and the change in the diurnal range in the skin temperature (not shown) is considerably increased throughout the year. Total runoff decreases by > 300 mm over the year indicating a decrease in moisture convergence into the area since, averaged over a number of years, the runoff must balance the difference between precipitation and evaporation.

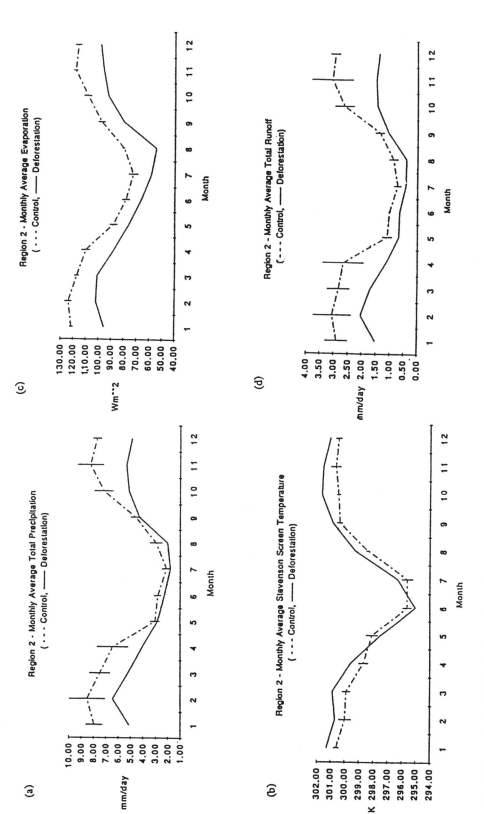

Fig. 4 Seasonal plots for Region 2 (southern Amazon) (five year averages from experiment and control simulations) showing twelve monthly values and ±2 standard errors on the control curve. Fields shown are: (a) total precipitation (mm d^{-1}), (b) Stevenson screen air temperature (K), (c) evaporation flux from surface (W m^{-2}), (d) total runoff (mm d^{-1}).

Region 1 lies to the north of the equator comprising seven $4.5° × 7.5°$ grid elements five of which are part of the deforestation land use change (Fig. 3). Precipitation decreases are similar to those in Region 2 in that the largest decrease in precipitation coincides roughly with the maxima in rainfall (Fig. 5(a)) which is similar to the situation in Region 2 where the largest decrease also took place in the rainy season (also November–March). This seasonality in rainfall decrease might have important implications for the possibility of reforestation or sustained agriculture in the Amazon Basin where exacerbation of the dry season is likely to be detrimental to vegetation regrowth and likely to increase the incidence of fires, especially in El Niño years. The results of this simulation indicate that the largest reduction in precipitation tends to coincide with the seasonal maxima thus, somewhat, reducing the impact for agriculture and forestry. Temperature changes are small. Only the Stevenson screen temperature (Fig. 5(b)) is increased following deforestation with both the ground surface and root zone temperatures being slightly lower in the deforested case than in the control.

Evaporation and total runoff are both decreased (Figs. 5(c),(d)) and, although the absorbed solar radiation is almost unchanged, the net radiation is also decreased by deforestation (not shown here).

The advantage of undertaking model experiments of more than a single year is that an assessment can be made of the statistical significance of the resulting response. As well as calculating confidence limits on the seasonal-cycle and time-series curves displayed in Figs. 4 and 5, our five-year control and five-year experiment have permitted the calculation of maps of Student's t values. It is recognized that this is an incomplete means of assessing the significance of detected responses (Chervin, 1980a,b, 1981; Preisendorfer and Barnett, 1983). Nonetheless, as physical interpretation will be the main means of assessing response, it was felt useful to employ a straightforward statistical test such as Student's t. Fig. 6 depicts differences (deforestation minus control) for selected fields for South America. In addition, there are smaller inset maps, shaded to show the regions of statistically significant increase (solid) and decrease (hatched) at the 95% level. Evaporation (Fig. 6(d)) and net surface radiation show the largest areal extent and seasonally most consistent response with both fields exhibiting statistically significant decreases following deforestation across the whole of the Amazon Basin. Temperatures (both Stevenson screen (b) and ground surface (a)) show statistically significant increases in January across most of the Basin but not in July (in agreement with the seasonal cycle curves shown in Figs. 4 and 5). The deforested region shows peak temperature increases of up to 2.8 °C (January) and 1.5 °C (July) for the ground surface temperature and around 2 °C (January) and 1.5 °C (July) in the case of the surface air temperature,

somewhat smaller than the maximum increases reported by Dickinson and Henderson-Sellers in 1988 which were, respectively, 4.5 °C (January) and 5.0 °C (July) in ground temperature and 1.8 °C and 3.2 °C in surface air temperature.

Total precipitation (Fig. 6(c)), total runoff and total cloud amount show decreased values but the decreases are not large enough to be statistically significant at the 95% level. In the CCM1-Oz simulations, runoff decreases over most of the deforested area in both months, although the decrease is more widespread in January (cf. Fig. 4(d)). The runoff differences differ in sign from those reported by Dickinson and Henderson-Sellers (1988) who found an increase in runoff although this result was, in part at least, due to their prescribed change in the exponent in the runoff formulation (see Table 2).

Since their prescription of increased runoff was deemed implausible, it has not been tested in a GCM simulation here, but comparative simulations were undertaken with BATS in an off-line mode (i.e. forced by prescribed atmospheric conditions). In two two-year simulations, the effect of Dickinson and Henderson-Sellers' (1988) runoff parameterization modifications was to increase the runoff in the deforested situation. The annual increase was ∼350 mm. With their parameterization, runoff always exceeded the tropical forest runoff whereas without the exponent change (as for the GCM experiments reported here) the predicted runoff was larger after deforestation early and late in the year but slightly lower than in the forest in the middle of the year. However, in the global simulations conducted here both Regions 1 and 2 show a decrease in total runoff throughout the year (Figs. 4(d) and 5(d)) indicating important feedbacks from the surface climate to the atmospheric circulation.

Precipitation is decreased in the deforested environment, particularly in January where the upper basin receives ∼135 mm less and the southeast part of the basin receives ∼320 mm less rainfall than in the control. In July, there is also a decrease in total precipitation, with an area of ∼175 mm decrease centred at the mouth of the Amazon. We find that the greatest decrease in precipitation occurs in the rainy period, in general agreement with Lean and Warrilow (1989). In contrast, Dickinson and Henderson-Sellers (1988) found a negligible change in rainfall and Shukla et al. (1990) found a tendency to exaggerate the dry conditions in mid-year. On the other hand, there is an area of statistically significant precipitation increase to the southwest of the deforested Amazon which is quite marked in July (Fig. 6(c)) which is easy to see in the (inset) Student's t map but much harder to discern in the difference field itself. This could, perhaps, be an indication of disturbances in the large scale moisture flow and possibly a climatic impact external to the deforested region. These topics are dealt with in the subsequent sections.

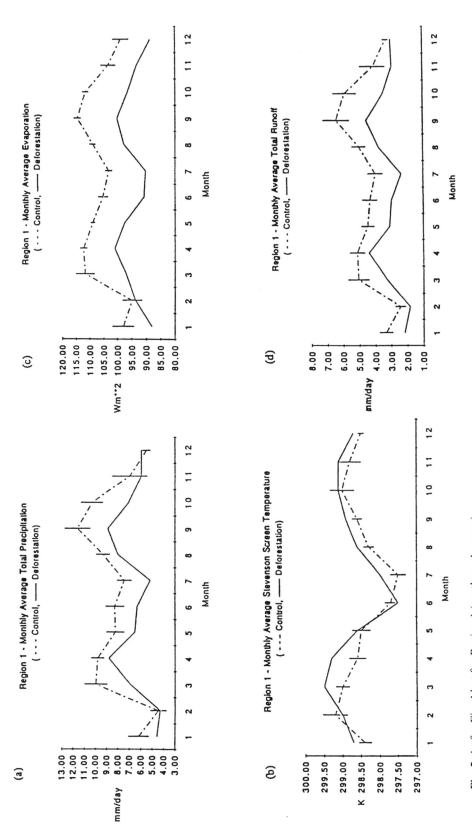

Fig. 5 As for Fig. 4 but for Region 1 (northern Amazon).

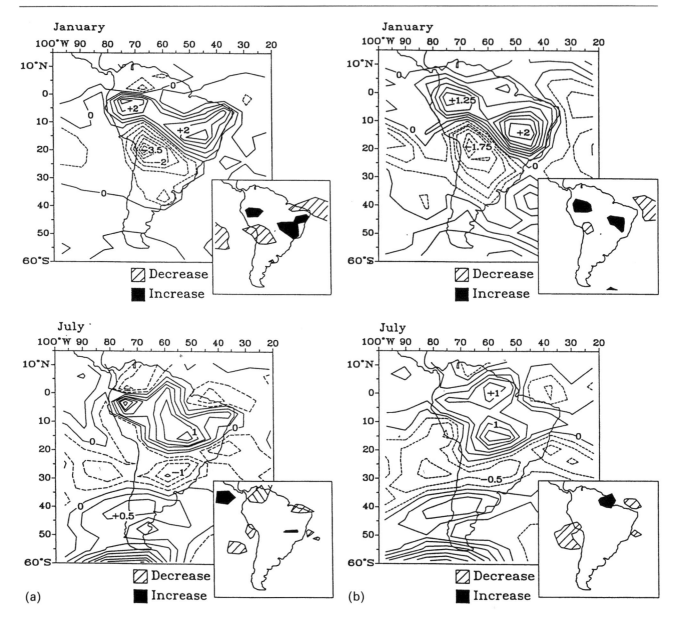

Fig. 6 Difference plots for South America for January (upper) and July (lower). In all cases the fields are derived from two sets of five-year integrations. Inset are the Student's *t* fields showing statistically significant (at 95% level) decreases (hatched) for values < − 2.306 and statistically significant increases (solid) for values > 2.306. Fields depicted are: (a) ground surface temperature (°C), (b) Stevenson screen temperature (°C), (c) total precipitation (mm), (d) evaporation (mm).

It is recognized that the Student's *t* test is not a fully satisfactory measure of statistical significance; in particular it does not test the significance of the response of geographical areas in which an apparently coherent pattern is associated (partially) with spatial autocorrelation (e.g. Hayashi, 1982; Storch, 1982; Livezey and Chen, 1983). Nonetheless, the Student's *t* maps for South America have indicated that changes in fields not generally discussed may have a more obvious signal than the climatic parameters of temperature and precipitation which are generally reported in the literature. In particular, we do not find statistical justification for a claim of a temperature rise in the middle of the year and little

or no statistical justification of the simulated decrease in precipitation.

2.2 Regional-scale impacts in SE Asia

The SE Asian region is important as the second focus for this experiment because it contains the next largest remaining area of contiguous tropical forest after the Amazon Basin and considerably different land-surface and climatic regimes. Also, since tropical deforestation is being investigated in the context of the global climate system, the teleconnections that may exist between the two regions should be investigated.

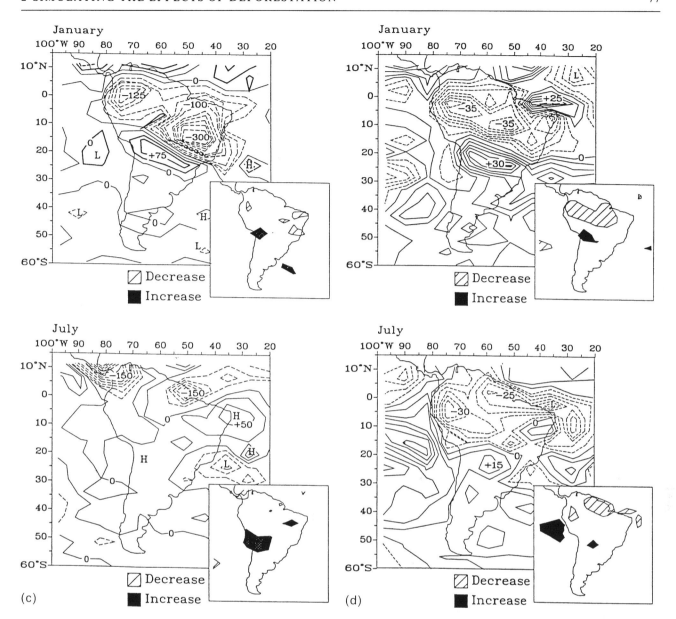

(c)

(d)

The area bounded by latitudes 20° N and 11° S and longitudes 95° E and 150° E containing the SE Asia land points (Region 3 in Fig. 3) has been spatially averaged. The averaging excludes the ocean points. It has been used to compute the natural variation of the climate for this region in the form of seasonal cycles showing ± 2 standard errors: approximately a 95% confidence interval.

Total precipitation (Fig. 7(a)), which is driven by the input of energy and moisture from the surrounding ocean (e.g. Meehl, 1992) is very little affected by the imposed deforestation: a decrease is seen in two of the wet season months (June and July) and marginal increases in January to April. The latter are not statistically significant.

The temperatures of the soil surface and root zone show very similar change with a decrease of about 0.5 to 1.0 K peaking in May, well outside the natural variability of the region. The Stevenson screen temperature (Fig. 7(b)) shows a less marked decrease of between 0 and 0.6 K with the same maximum in May and a similar seasonal cycle. The diurnal skin temperature range (not shown) increases by between 1.2 and 2.5 K for all months and well exceeds the confidence interval of ± 2 standard errors. Evaporation shows an overall decrease of between about 5 and 15 W m^{-2} for the spatially averaged region in the dry and wet seasons respectively (Fig. 7(c)), as does the surface absorbed solar radiation and the net surface radiation (not shown). The total runoff pattern is very similar to the precipitation distribution but with even smaller differences. However, the surface runoff (Fig. 7(d)) differs from this (unlike Regions 1 and 2 where surface and total runoff differences are very similar). Surface runoff increases throughout the year following deforestation.

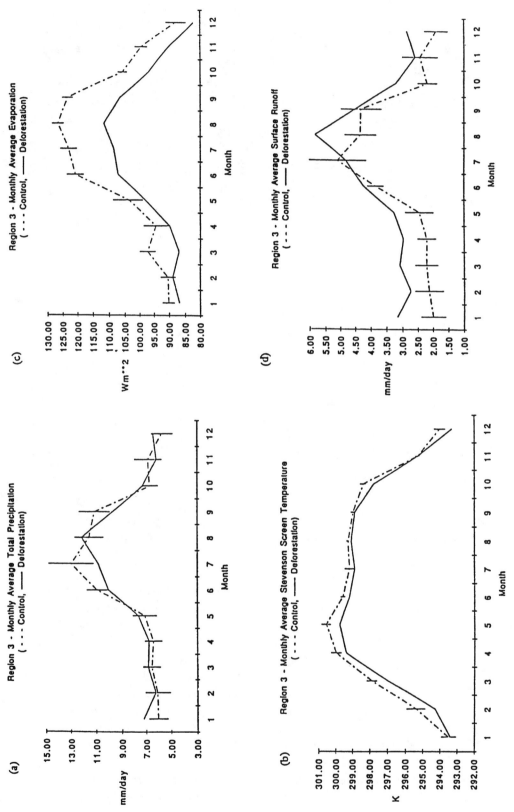

Fig. 7 As for Fig. 4 except for SE Asia (Region 3) and part (d) is surface (not total) runoff.

The dry season (January) values of soil surface temperature differences (Fig. 8(a)) show an increase of approximately 0.8 K in the south of the continental part of this region after deforestation. However, the temperature changes over the other deforested land points are negligible. There is also a 1 K decrease in the soil surface temperature to the northwest of continental SE Asia and a 1.2 K increase to the northeast but neither are statistically significant. In July (the wet season) (Fig. 8(a)) there is a decrease in soil surface temperature of 0.9 K over the continental part of the region, a 0.7 K decrease over the island of Borneo and a 1.6 K decrease over North Australia, all of which are significant changes.

Total precipitation (Fig. 8(b)) is difficult to investigate in the context of deforestation in this region, because a vast amount of rainfall is received throughout the year, particularly in the wet season, as a result of the strong forcing of rainfall by such events as orographic uplift and its continual 'fuelling' from the adjacent tropical ocean. The areas of tropical forest play a lesser role in producing the high rainfall of this area than in the Amazon region. Consequently, many of the changes in precipitation that have resulted from the deforestation experiment are not statistically significant as they do not exceed the large natural variation of this region.

In the dry season (January), there is a 140 mm increase in rainfall to the south of the mainland, indicating a northward shift in the maximum shown in the control plot. There is also an increase to the east of New Guinea of about 100 mm, but there are no important changes over the deforested regions. Fig. 8(b) also shows that in the wet season (July), there is a general decrease over the deforested continental part of SE Asia of about 50–75 mm and an increase of about 110 mm to the east of Borneo. North of New Guinea there is a large increase of 150 mm and a decrease further south of about 75 mm. There is also a large decrease of 180 mm in the precipitation maximum to the northwest of continental SE Asia. Although some specific point values have fluctuated with the deforested case there is no change to the basic pattern of rainfall and, as the Student's t tests show, few of these changes are statistically significant.

Fig. 8(c) shows evaporation differences and their statistical significance. There are large changes in the deforested areas which are significant. There is a 24 W m^{-2} decrease in evaporation over Borneo in the dry season and a 12 W m^{-2} decrease over New Guinea. Of less significance is the decrease in evaporation over the continental part of SE Asia of 4 to 8 W m^{-2} and the increase of 10 W m^{-2} to the east of Borneo. On the whole, the pattern of evaporation has changed little following deforestation. In July, the overall pattern has also remained the same but the three main deforested land points show considerably significant changes in evaporation. There is a decrease of about 20 W m^{-2} over the island of Borneo and a slightly smaller decrease over New

Guinea. There is also a 20 W m^{-2} decrease over continental SE Asia.

The changes in net radiation (Fig. 8(d)) for January show no alteration to the seasonal pattern but show significant changes to the values over the land points. There is a decrease of approximately 2.5 to 7.5 W m^{-2} over continental SE Asia and a much larger decrease of about 17 W m^{-2} over Borneo and 11.5 W m^{-2} over New Guinea. There is also a statistically significant area of increased net radiation over the northern Philippines of about 13 W m^{-2} which is difficult to explain. The wet season (July) shows similar areas of statistically significant changes, with a 10 to 18 W m^{-2} decrease in net radiation over the continental part of the region, a decrease of 17.5 W m^{-2} over Borneo and a 13.5 W m^{-2} decrease over New Guinea.

2.3 Time scale of response to tropical deforestation

An important question which has never been adequately addressed is the issue of the length of time required for a coupled land-surface scheme and climate model to adjust to an imposed disturbance such as large scale land use change. Stand-alone simulations with the BATS land-surface scheme show that this model requires at least 3–4 months before it adjusts to initialization or prescribed changes (Pitman et al., 1990). Even after this adjustment time, the lower soil temperatures and moisture contents are not in equilibrium with the prescribed atmospheric forcing and can take 1–2 years to stabilize completely. Thus, during the first year following large scale deforestation, the large scale climate response might be somewhat transient rather than fully equilibrated. Indeed, the precise timing of full equilibration is very difficult in the GCM context since the hypothesis underlying all these experiments is that the prescribed land use change is likely to disturb the surface climate which, in turn, may perturb the atmosphere overlying it and, indeed, other regions. If this is correct, it may go some way to explaining the differences amongst the four previous GCM simulations of the impact of tropical deforestation. Of these, the experiments of Dickinson and Henderson-Sellers (1988) and Shukla et al. (1990)/Nobre et al. (1991) were run only for 13 and 12 months respectively even though both included a complex land-surface scheme (see Table 1). So the impacts reported might have been, at least in part, a transient response to the imposed land use change although this is only one of a number of sources of discrepancy.

Fig. 9 is an attempt to assess the length of time required for the large scale climate to adjust to a large scale land use change. The figure shows time series for the latter 5 years of the 6-year deforestation experiment for Region 2 (Southern Amazon). The plots show some, weak, trends but, more importantly, they also indicate that there may be a 12–18

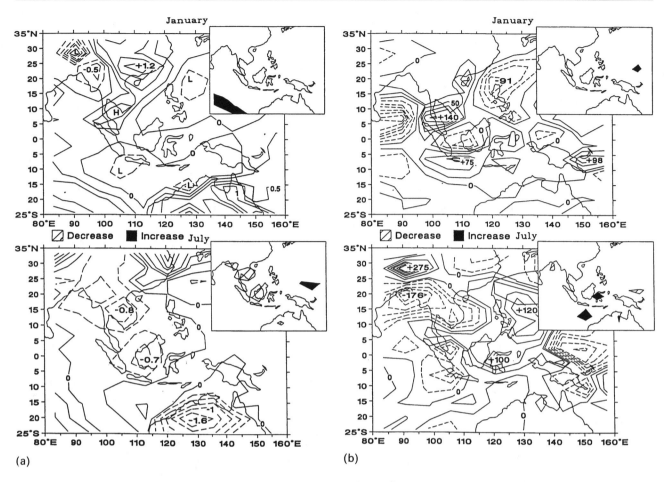

Fig. 8 As for Fig. 6 except for SE Asia and for the fields: (a) ground surface temperature (°C), (b) total precipitation (mm), (c) evaporation (W m^{-2}), (d) net radiation at the surface (W m^{-2}).

month period of adjustment following the land-surface change. Note especially that the soil moisture in the root zone (defined as the upper two soil layers in the BATS scheme) takes at least 3 years to equilibrate (Fig. 9(e)). Indeed, the difference between the control and deforestation experiments for this parameter exhibits an almost monotonic decline over the whole of the 6-year period. This time scale of response reflects the length of time required by the deep soil layer to equilibrate with altered surface conditions and a modified climate because the root zone moisture interacts with this deeper moisture store. The deep soil zone was introduced into the BATS code subsequent to the Dickinson and Henderson-Sellers (1988) experiment to make BATS' behaviour more similar to the SiB scheme. Thus this long-time-scale response may not have greatly impacted the results of Dickinson and Henderson-Sellers (1988) but is likely to be affecting those of Shukla et al. (1990).

The analysis described here has been based on the latter five years of the six-year experiments to avoid some of these discontinuities. Even in this five-year period, however, the transient effects of the instantaneous land use change can still

be discerned. Fig. 9 shows five climatic variables which seem to continue to show a non-stationary response to deforestation beyond the first 24–30 months viz. absorbed solar radiation, diurnal range in skin temperature, sensible heat flux, total cloud amount and root soil moisture. Fig. 9(d) shows the total cloud amount which, although not exhibiting a trend in total amount, does exhibit an apparent increase in variability with maximum monthly cloud amounts increasing over the period while minimum amounts decrease.

Overall, it seems reasonable to draw two conclusions from Fig. 9. Assessment of the impact of tropical deforestation after only one year is, at best, incomplete since it is very likely that the local climate has not fully equilibrated to the imposed disturbance. On the other hand, a five-year period seems to be long enough for equilibration to have been adequately achieved. In short sensitivity tests (less than 12 months) 'changes' in climatic parameters that exhibit rapid and, subsequently, stable responses (e.g. evaporation) will be more meaningful than those of the slow-response or noisy fields (e.g. precipitation and soil moisture).

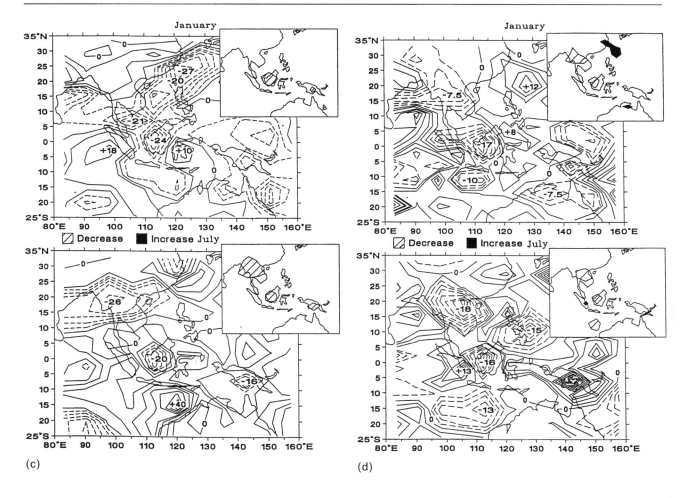

(c)

(d)

3. LARGE SCALE IMPACTS OF TROPICAL DEFORESTATION

3.1 Walker and Hadley circulation changes

It has been hypothesized that large scale land-surface disturbance, especially if it occurs coincidentally in time in more than one geographical location, might prompt circulation changes beyond the area of prescribed change. The only previous GCM experiment which was for a long enough time-period to permit this type of investigation (Henderson-Sellers and Gornitz, 1984) did not detect changes 'external' to the deforested area. As this investigation differs in a number of important ways from this earlier study (viz. higher spatial resolution GCM, more complex land-surface scheme, better large scale climate prediction for the present-day) it was thought useful to investigate the possibility of larger scale changes once again.

One anticipated outcome of tropical deforestation is a reduction in the vertical ascent over the deforested region(s) caused by the imposed increase in surface albedo and hence net loss of energy to the column and also by the decrease in net surface radiation prompted by higher surface temperatures caused by the smaller turbulent exchanges and, finally, by the reduction in evaporation resulting from the imposed

decreases in canopy extent and vegetation roughness length. This, in turn, might affect the components of either the Walker or the Hadley circulations in the area and, more importantly, the effects of the deforestation in the two regions might interact: dampening or, perhaps, re-inforcing the 'external' disturbances they prompt. Here we assess these possibilities first by considering the changes in the vertical velocity (Pa s^{-1}) over the two deforested regions. Here latitudinal cross-sections are used to capture the cells of the Walker circulation while the two longitudinal cross-sections are to depict the latitudinal movement of the ITCZ in the two regions of the deforestation: Amazon Basin and SE Asia.

Fig. 10 shows the Walker circulation for the control (a) and (b), the deforestation experiment (c) and (d) for January and July respectively. The control experiment shows the cells of the Walker circulation very clearly in both months. In both seasons there is large-scale ascent over the Amazon Basin ($\sim 50°$ W), over SE Asia ($\sim 100°$ E) and over the western Pacific ($\sim 140°$ E$-180°$ E). In January (Fig. 10(a)) there is additionally an ascending limb over tropical Africa ($\sim 25°$ E) but this does not appear in the July cross-section (Fig. 10(b)). Also, ascent over the Amazon is significantly weaker in July than in January because the ITCZ lies well to the north of the Basin in July (see inset map, Fig. 10(e)).

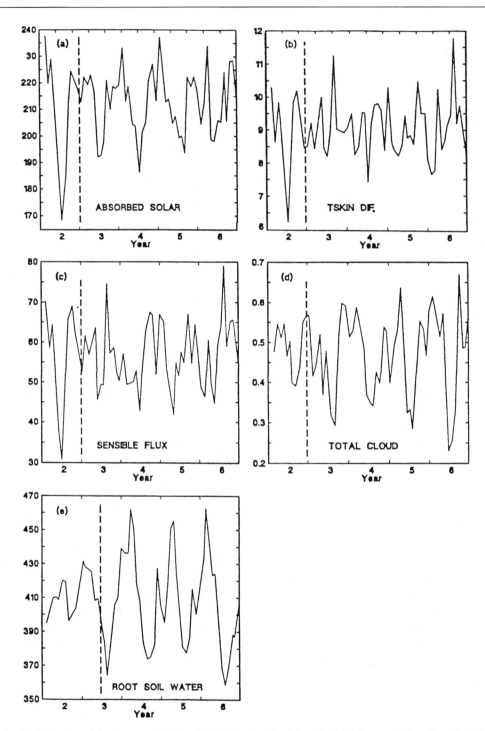

Fig. 9 Time series for latter five of the six-year deforestation illustrating the delayed establishment of the effect of the instantaneous deforestation. Plots are regionally averaged values from the experiment for Region 2 (southern Amazon). The dashed vertical line is at the end of the second year of integration except for root zone moisture where it is at the end of 2.5 years of integration. Fields shown are: (a) absorbed solar radiation at the surface (W m^{-2}), (b) diurnal range in skin temperature (K), (c) sensible heat flux from the surface (W m^{-2}), (d) total cloud amount (fraction), (e) root-zone soil water content at 00Z (mm).

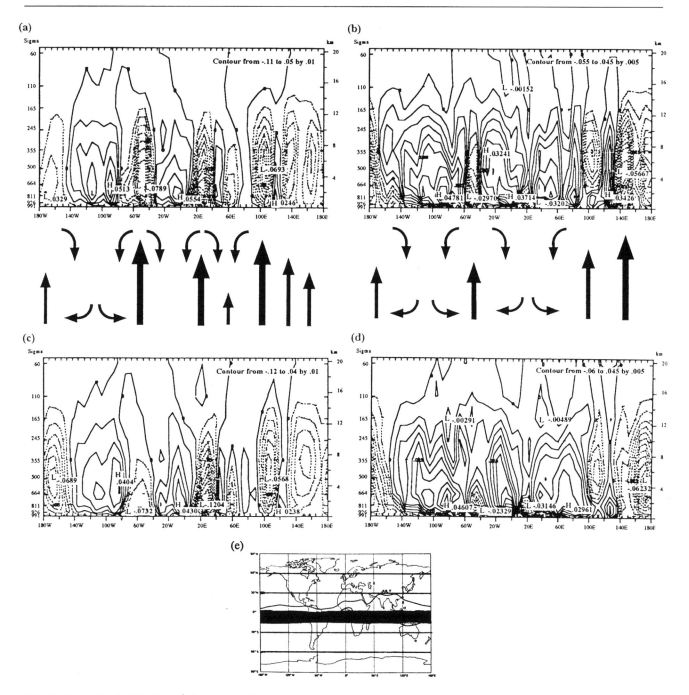

Fig. 10 Vertical velocities (Pa s^{-1}) in the latitudinal cross-section from 180° W to 180° E and from 2.2° N to 15.6° S for five-year means. (a) January: control, (b) July: control, (c) January: deforestation, (d) July: deforestation, (e) location map showing the cross-section represented.

Difference plots (not shown) are rather difficult to interpret as the sign of the differences depends upon the signs of the vertical velocities in the experiment and control but are a useful pointer to the regions of largest change. These seem to be the Amazon Basin ($\sim 50°$ W) in January and for the eastern edge of SE Asia in July ($\sim 140°$ E).

The effects of deforestation are easier to assess by direct comparison of the control and experiment vertical velocities but note that the ranges (but not the contour interval) differ slightly. In both seasons the ascent over the Amazon is

considerably diminished in the deforestation experiment. The effect is greater in January because the ascent is very much stronger in the January control climate. This result is in agreement with the finding of Henderson-Sellers and Gornitz (1984). The impact of deforestation on the Walker circulation is essentially to remove the Atlantic and east Pacific cells in July so that descent predominates from $\sim 140°$ W to $\sim 100°$ E.

Fig. 11 depicts the Hadley circulation as captured by a cross-section through the atmosphere extending from 45° W

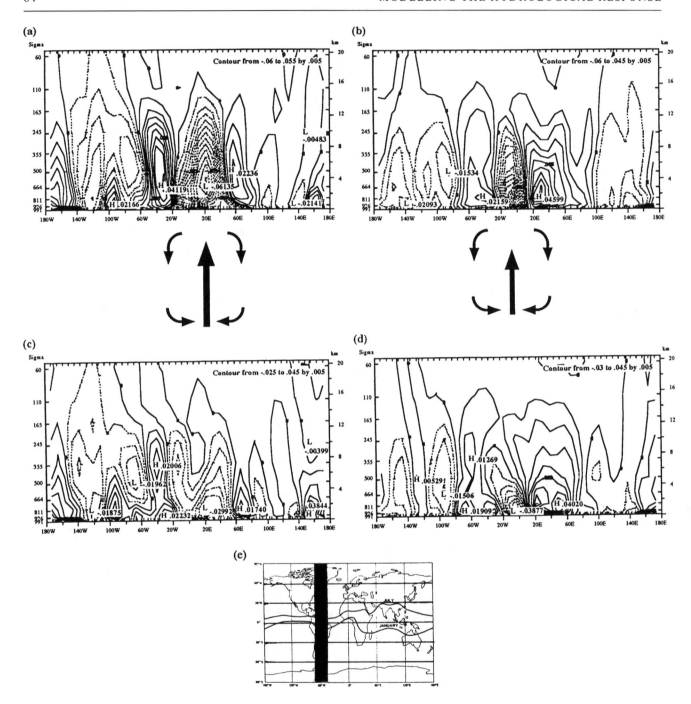

Fig. 11 As for Fig. 10 except showing a longitudinal cross-section from 90° N to 90° S and from 45° W to 75° W (i.e. including the Amazon Basin).

to 75° W (i.e. including the Amazon Basin). In the control simulation (Figs. 11(a) and (b)) the ITCZ is clearly seen at 10° S in January and about 5° N in July. After deforestation (Figs. 11(c) and (d)) both the upwelling in the ITCZ and the descent in both subtropical regions is weakened although the effects do not extend outside about 40° N to 40° S. Unlike the impact on the Walker circulation, the effects seem to be of similar magnitude in both January and July.

Vertical velocities have also been assessed (not shown here) in a latitude band from 90° E to 112° E, which was selected to encompass the mainland areas of SE Asia where deforestation is imposed. The control case shows the position of the ITCZ: on the equator in January and centred at about 12° N in July. The effects of deforestation are very much smaller in this latitude band which excludes the South American forest. The vertical ascent in the ITCZ is slightly diminished in both seasons but there are no other discernible effects.

Overall, the impact on the general circulation of the atmosphere can be detected in the cells of the Walker circulation in both January and July and results in diminished ascent in the ITCZ over both deforested areas. There is also a decrease in the circulation strength of the Hadley cell over South and Central America. The effects are considerably greater over South America than over SE Asia.

3.2 Large scale hydrological responses to deforestation

Shukla et al. (1990) detected decreases in the low-level, basin scale convergence as a result of their deforestation experiment using the NMC GCM and the SiB land-surface scheme. This result was contrary to previous suggestions that feedbacks at the land-surface might prompt increased moisture convergence to 'compensate' for decreased evaporation and increased runoff following deforestation.

In this experiment, which was designed to permit examination of this possible climatic impact, it seems that there are similar changes since (i) the vertical ascent over the Basin is greatly reduced by deforestation and (ii) the difference between precipitation and evaporation in Region 2 changes from control to deforested climates. Table 1 summarizes the differences between precipitation and evaporation found between the control and deforestation experiments. In the case of the annual mean, we find the moisture divergence ($=E-P$) in the deforested regime to be only 55% of the control case whereas Shukla et al. (1990) found it to be $\sim 82\%$. Month-by-month analysis indicates that the greatest decrease in convergence occurs in the period November to February (the rainy season in Region 2).

One method of assessing changes to the basin-wide circulation is to plot differences in the divergence fields between the deforested experiment and the control. Fig. 12 illustrates these differences in divergence for the lower (811 σ) troposphere for January and July. In January (Fig. 12(a)) the lower-level flow shows a large area of decreased convergence (positive contours of increased divergence). There is a matching area of decreased divergence aloft. These differences are in keeping with the considerable weakening in ascent over the Amazon (described in the previous section) and observed changes in the near-surface wind flow in January. There are similar, but weaker and less spatially extensive, features in the July fields of divergence differences (Fig. 12(b)). It is also interesting to note that the region immediately to the south of the Amazon Basin undergoes similar changes in divergence but of the opposite sign in both seasons indicating that near-surface convergence is increasing in this area following the deforestation of the Amazon.

The Amazon Basin undergoes changes following deforestation to its circulation pattern of considerable magnitude in January and of similar sign but lesser magnitude and extent

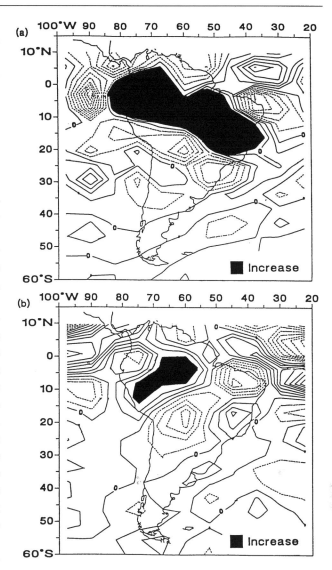

Fig. 12 Differences (deforestation – control) in divergence fields at 811 σ for (a) January and (b) July.

in July. Specifically the changes comprise: (i) decreased ascent (and in July almost complete removal of this ascending limb within the Walker circulation); (ii) decreased surface convergence and decreased divergence aloft (especially in January when the ITCZ is located over the Amazon); and (iii) increased surface convergence to the south of the deforested region. These changes seem to extend beyond the deforested area especially in their effect upon the cellular pattern of the Walker circulation and the outflow from the deforested region to the area to its south.

This simulation experiment was also designed to investigate the hypothesis that disturbances to the circulation pattern beyond the deforested area particularly to the south of the Amazon region (between 20° S–30° S) might occur after deforestation. This hypothesis was prompted by alterations to the local climate to the south of the deforested area (clearer in January than in July) which can be identified in the results of Dickinson and Henderson-Sellers (1988). The model results show significant changes in climatic parameters of opposite sign and of slightly smaller magnitude and smaller areal extent to those occurring in the deforested region itself. Specifically, temperatures decrease, evaporation increases and precipitation increases (e.g. Figs. 6(a)–(d)). Ground surface temperatures are lower by around −1 °C (July) and −4 °C (January) and precipitation is increased by between 23 mm (July) and 90 mm (January) (Fig. 6(c)).

The major cause of disturbances outside the deforested region seems to be the reduced ascent in the ITCZ (e.g. Fig. 10). In January, the equatorial trough lies well to the south of the equator and its effects, combined with the continental heat low, are to generate strong ascent in the control climate. After deforestation, however, this ascent and the associated surface convergence are weakened over the basin but there is increased surface convergence to the south. In this season, evaporation is increased and the surface temperatures are lower, probably as a direct consequence of the larger evaporative flux.

In July, the weakening of both the convergence into the ITCZ and the ascent occurs to the north of the deforested region. The main consequence of this is a strengthening of the (already strong) easterly flow throughout the Amazon region. To the south of the deforested area, there is again increased surface convergence, although in this season, the air flow is primarily from the east (Atlantic), rather than from the north (the deforested area) as in January and thus produces a geographically less coherent response south of the deforested region.

If the climatic changes indicated here, viz. lower temperatures and greater rainfall, were corroborated in other studies, it is possible that these impacts could be seen as benefitting agriculture lying to the South of the Amazon. Such arguments would be unwise in view of the considerable uncertainties surrounding simulations of the climatic impacts of land use change and that the changes occur mostly during the wet season when rainfall is ample.

4. ASSESSMENT OF THE SIMULATED IMPACTS OF LARGE SCALE LAND USE CHANGE

This is the first tropical deforestation simulation with a global climate model which has explored the impacts of simultaneous deforestation in both the Amazon and in SE Asia. Previous GCM experiments prompted a series of hypotheses which has been examined here. These were, in decreasing order of likelihood: (i) surface climatic changes where deforestation was prescribed; (ii) disturbances to the

vertical motion over and horizontal advection into the deforested areas; (iii) a climatic response external to the deforested Amazon (specifically to the south).

We find a larger hydrological response, in the form of a large decrease in basin scale moisture convergence, than all previous experiments, i.e., the evaporative flux decrease (which we find to be statistically significant through the year) is far outweighed by the decrease in precipitation. On the other hand, we find smaller temperature increases in the deforested Amazon than did any earlier study except Henderson-Sellers and Gornitz (1984). Indeed, in the northern Amazon Basin, temperatures generally decrease as also happens throughout SE Asia.

These differences from earlier experiments could be due to a number of causes including the specification of land-surface characteristics for both the forest and the deforested scenarios. Another factor is the length of the simulations and the number of soil layers in the land-surface scheme. Our experiment was undertaken for long enough for the climate to become equilibrated to the land-use change (at least 1–2 years) whereas all the previous experiments except that of Henderson-Sellers and Gornitz (1984) were conducted only for about 12 months.

The results reported here prompt a number of questions including: (1) how applicable are the parameterizations 'stolen' from microhydrology to the coarse resolution GCMs to which they are now applied (cf. Fig. 1); (2) why the experiment described here produces larger hydrological responses (runoff decrease and larger decrease in moisture convergence) than any previous simulations of Amazonian deforestation; (3) why the precipitation decreases in the Amazon (north and south of the basin) occur primarily in the rainy season (∼30% decrease versus ∼10% in the dry season); (4) why surface temperatures do not increase very much and, indeed, decrease early in the dry season. The major difference in the experiment reported here is in the roughness length prescribed for the deforested scrub grassland. Following Wieringa's (1992) recent review, we selected a roughness length of 0.2 m which is larger than the previous values which ranged from 0.01 m (Henderson-Sellers and Gornitz, 1984) to 0.08 (Shukla et al., 1990).

The mechanism likely to prompt a change in moisture convergence is a Charney-type disturbance in which less energy is absorbed, causing a decrease in vertical ascent and, thus, a reduction in convergence. Such a change would seem to owe its origin primarily to an albedo increase (e.g. Charney, 1975) although it is possible to argue that a decrease in net surface radiation could also be the result of an imposed roughness length decrease which prompted a decrease in sensible and latent fluxes and, hence, an increase in surface temperature. Overall, it appears that the results reported here are interpretable in terms of the albedo decrease combined with a smaller roughness decrease than in other experiments. Together these prescribed changes to the surface characteristics could explain the large decrease in moisture convergence combined with the negligible increases in surface temperatures.

Tropical deforestation in the Amazon differs in impact from that for SE Asia primarily in that in S America there is a significant decrease in near-surface moisture convergence, which prompts associated circulation and surface runoff changes, whereas convergence changes in SE Asia were not identified. This response is explicable in terms of the relative changes in the two components of the total water vapour convergence. In both areas evaporation decreases and hence near-surface specific humidity decreases. In the Amazon, this results, directly, in a decrease in moisture convergence. In SE Asia, however, the horizontal distances between ocean and deforested land are so small (relative to the Amazon Basin) that the gradient of specific humidity becomes large enough to balance the induced moisture convergence decrease.

In addition to these large scale changes, it has also been shown that there are alterations to the basin scale hydrology in the Amazon region and that detectable rainfall increases occur in the climate to the south of the deforested area in South America. In addition, we find that the precipitation decreases which occur in all deforested regions affect the rainy seasons more than the dry seasons (minimizing exacerbation of drought conditions in deforested areas). Such climatic changes, if corroborated by other experiments, could have implications for forest and agricultural sustainability. The question of teleconnections to adjacent regions and across the Pacific needs to be resolved by means of further GCM sensitivity studies.

ACKNOWLEDGEMENTS

This research was supported in part by the Model Evaluation Consortium for Climate Assessment, the Australian Research Council, the Australian Department of Arts, Sport, Environment and Territories. The Australian authors wish to thank Dr Warren Washington for hospitality in the Climate and Global Dynamics Division of NCAR during the development of the research and Dr Robert Dickinson for grants supporting the research undertaken at the University of Arizona. This is Contribution no 92/12 of the Climatic Impacts Centre.

REFERENCES

AUSTRALIAN RAINFOREST MEMORANDUM, 1991, Rainforest Information Centre, Lismore, NSW, Australia, 30pp

AVISSAR, R., 1993, A method to bridge the gap between microscale and regional-scale hydrological processes, this volume

BECKER, A., 1993, Determinism and statistical distribution principles in large-scale hydrological modelling, this volume

BOLIN, B., 1977, Changes of land biota and their importance for the carbon cycle, Science, 196, 613–621

CHARNEY, J.G., 1975, Dynamics of deserts and drought in the Sahel, Quart. J. Roy. Meteor. Soc., 101, 193–202

CHERVIN, R.M., 1980a, Estimates of first- and second-moment climate statistics in GCM simulated climate ensembles, J. Atmos. Sci., 37, 1889–1902

CHERVIN, R.M., 1980b, On the simulation of climate and climate change, with general circulation models, J. Atmos. Sci., 37, 1903–1913

CHERVIN, R.M., 1981, On the comparison of observed and GCM simulated climate ensembles, J. Atmos. Sci., 38, 885–901

DICKINSON, R.E., 1989, Implications of tropical deforestation for climate: a comparison of model and observational descriptions of surface energy and hydrological balance, Phil. Trans. Roy. Soc. Lond. B, 324, 423–431

DICKINSON, R.E. and HENDERSON-SELLERS, A., 1988, Modelling tropical deforestation: a study of GCM land-surface parameterizations, Quart. J. Roy. Meteor. Soc., 114(B), 439–462

DICKINSON, R.E., HENDERSON-SELLERS, A., KENNEDY, P.J. and WILSON, M.F., 1986, Biosphere-Atmosphere Transfer Scheme (BATS) for the NCAR Community Climate Model, National Center for Atmospheric Research, Boulder, BO, Tech Note/TN-275 + STR

HAMPICKE, V., 1980, The role of the biosphere, pp149–167 in Interactions of Energy and Climate (eds. W. Bach, J. Pankrath and J. Williams) D. Reidel, Dordrecht, Holland

HAYASHI, Y., 1982, Confidence intervals of a climatic signal, J. Atmos. Sci. 39, 1895–1905

HENDERSON-SELLERS, A., 1992, Climate modelling of tropical deforestation: can we improve the current parameterizations?, Quart. J. Roy. Meteor. Soc. (accepted for publication)

HENDERSON-SELLERS, A. and GORNITZ, V., 1984, Possible climatic impacts of land cover transformations, with particular emphasis on tropical deforestation, Climatic Change, 6, 231–258

KINTER, J.L., SHUKLA, J., MARX, L., and SCHNEIDER, E.K., 1988. A simulation of the winter and summer circulations with the NMC global spectral model. J. Atmos. Sci. 45:2486–2522

LEAN, J. and WARRILOW, D.A., 1989, Simulation of the regional climatic impact of Amazon deforestation, Nature, 342, 411–413

LIVEZEY, R.E. and CHEN, W.Y., 1983, Statistical field significance and its determination by Monte Carlo techniques, Mon. Weath. Rev., 111, 46–59

MEEHL, G.A., 1992, Effect of tropical topography on global climate, Ann. Rev. Earth Planet. Sci., 20, 85–112

NOBRE, C.A., SELLERS, P.J. and SHUKLA, J., 1991, Amazonian deforestation and regional climate change, J. Clim., 4, 957–988

PITMAN, A.J., HENDERSON-SELLERS, A. and YANG, Z-L., 1990, Sensitivity of regional climates to localized precipitation in global models, Nature, 346, 734–737

POTTER, G.L., ELLSAESSER, H.W., MACCRACKEN, M.C. and ELLIS, J.S., 1981, Albedo change by man: test of climatic effects, Nature, 291, 47–50

PREISENDORFER, R.W. and BARNETT, T.P., 1983, Numerical model-reality intercomparison tests using small-sample statistics, J. Atmos. Sci., 40, 1884–1896

SAGAN, C., TOON, O.B. and POLLACK, J.B., 1979, Anthropogenic albedo changes and the earth's climate, Science, 206, 1363–1368

SELA, J.G., 1980, Spectral modeling at the N.M.C., Mon. Wea. Rev., 108, 1279–1292.

SELLERS, P.J., MINTZ, Y., SUD, Y.C. and DALCHER, A., 1986, A simple biosphere model (SiB) for use within general circulation models, J. Atmos. Sci., 43, 505–531.

SHUKLA, J., 1987, General circulation modelling and the tropics, pp.409–458, The Geophysiology of Amazonia (ed. R.E. Dickinson), John Wiley & Sons, NY

SHUKLA, J. and MINTZ, Y., 1982, Influence of land-surface evapotranspiration on the Earth's climate, Science, 215, 1498–1501

SHUKLA, J., NOBRE, C. and SELLERS, P.J., 1990, Amazon deforestation and climate change, Science, 247, 1322–1325

SHUTTLEWORTH, W.J., 1988, Evaporation from Amazonian rain forest, Proc. Roy. Soc. Lond. B 233, 321–346

SHUTTLEWORTH, J.W. and DICKINSON, R.E., 1989, Comments

on 'Modelling tropical deforestation: a study of GCM, land-surface parameterizations' by R.E. Dickinson and A. Henderson-Sellers, Quart. J. Roy. Meteor. Soc., 115, 1177–1179

SLINGO, A., 1989, A GCM parameterization for the shortwave radiative properties of water clouds, J. Atmos. Sci., 46, 1419–1427

STORCH, H.V., 1982, A remark on Chervin-Schneider's algorithm to test significance of climate experiments with GCMs, J. Atmos. Sci., 39, 187–189

WIERINGA, J., 1993, Representative roughness parameters for homogeneous terrain, Boundary-Layer Meteorol., 63, 323–363

WILLIAMSON, G.S. and WILLIAMSON, D.L., 1987, Circulation statistics from seasonal and perpetual January and July simulations with the NCAR Community Climate Model (CCM1): R15, National Center for Atmospheric Research, Technical Note, NCAR/TN-302 + STR, 199pp

WILLIAMSON, D.L., KIEHL, J.T., RAMANATHAN, V., DICKINSON, R.E. and HACK, J.J., 1987, Description of the NCAR Community Climate Model (CCM1), National Center for Atmospheric Research, Technical Note, NCAR/TN-285 + STR, 112pp

WILSON, M.F., 1984, Construction and use of land surface information in a general circulation climate model, Ph.D. Thesis, Univ. of Liverpool, Liverpool, UK

WILSON, M.F. and HENDERSON-SELLERS, A., 1985, A global archive of land cover and soils data for use in general circulation climate models, J. Clim., 5, 119–143

WOODWELL, G.M., WHITTAKER, R.H., REINER, W.A., LIKENS, G.E., DETWICKE, C.C. and BOTKIN, D.B., 1978, The biota and the world carbon budget, Science, 199, 141–146

An approach to represent mesoscale (subgrid-scale) fluxes in GCMs demonstrated with simulations of local deforestation in Amazonia

R. AVISSAR and F. CHEN

Department of Meteorology and Physical Oceanography,

Cook Campus, Rutgers University,

New Brunswick, NJ 08903, USA

ABSTRACT Mesoscale circulations generated by landscape discontinuities (i.e. 'sea-breeze' like circulations) are likely to have a significant impact on the hydrological cycle, the climate, and the weather. However, these processes are not represented in large-scale atmospheric models (e.g. general circulation models (GCMs)), which have an inappropriate grid-scale resolution. Assuming that atmospheric variables can be separated into large-scale, mesoscale, and turbulent-scale, Avissar and Chen (1993) developed a set of prognostic equations applicable in GCMs for momentum, temperature, moisture, and any other gaseous or aerosol material, which include both mesoscale and turbulent fluxes. They suggested using the mean mesoscale kinetic energy (MKE) per unit of mass for parametrizing these mesoscale fluxes in such models. In the present study, which complements the work of Avissar and Chen (1993), we simulated the atmospheric planetary boundary layer (PBL) that develops above a locally deforested area of the Amazonian region, to illustrate the relationships that exist: (i) between the diurnal variation of MKE and the diurnal variation of mesoscale latent and sensible heat fluxes; and (ii) between landscape discontinuities resulting from horizontal gradients of moisture at the ground surface and MKE. We compared MKE with turbulence kinetic energy (TKE) to emphasize the magnitude of mesoscale processes, as compared to turbulent processes. This analysis illustrates the potential use of MKE to bridge between landscape discontinuities and mesoscale fluxes and, therefore, to parametrize mesoscale fluxes generated by such subgrid-scale landscape discontinuities in GCMs and other large-scale atmospheric models. Furthermore, our simulations indicate that local deforestation tends to increase vertical moisture heat flux in the PBL by the combined effect of turbulence and mesoscale circulation. As a result, local deforestation seems to increase cloudiness and precipitation, in contradiction to the prediction of current GCMs.

1. INTRODUCTION

It is widely agreed upon that the parametrization of the earth's surface is one of the more important aspects of climate modeling, since this surface absorbs over 70% of the energy absorbed into the climate system, and many physical, chemical, and biological processes take place there. Of particular importance are the exchanges of mass (notably water), momentum, and energy between the surface and the atmosphere (Avissar and Verstraete, 1990).

In current state-of-the-art general circulation models (GCMs), land-surface parametrizations are based on the concept of the 'big leaf', which implies that the land represented in each grid element of the model is homogeneously covered by a big leaf. However, at the resolvable scale of GCMs, continental surfaces are very heterogeneous. This can be readily appreciated, for instance, by examining maps of soil, vegetation, topography, or land use patterns. Thus, recently, parametrizations based on a statistical-dynamical approach have been suggested (Avissar and Pielke, 1989; Entekhabi and Eagleson, 1989; Avissar, 1991; Famiglietti and Wood, 1991; Avissar, 1992). With this approach, probability density functions (pdfs) are used to represent the variability of the various characteristics of the soil-plant-atmosphere system that affect the input and redistribution of energy and water at the land surface. Collins and Avissar (1994) and Li and Avissar (1994) showed that stomatal conductance, leaf area index, and surface roughness are the most important characteristics to be represented by pdfs, and that using the mean value of these characteristics rather than their pdf could introduce significant errors in the calculation of surface energy fluxes.

But extended landscape heterogeneities that result, for instance, from the juxtaposition of land and bodies of water or bare-soil and vegetated areas are likely to produce 'sea-breeze' like mesoscale atmospheric circulations (e.g. Mahfouf et al., 1987; Segal et al., 1988; Avissar, 1991; Mascart et al., 1991). These circulations may have a significant impact on various atmospheric processes, e.g. cloud formation, aerosols, and gas transport, etc. For instance, Avissar (1991) discussed the possible effects of developing agriculture in arid regions and in deforested tropical forests on convective cloud formation. He concluded that these important processes, which develop at a smaller scale than the resolvable scale of GCMs, are likely to affect significantly the hydrological cycle, the climate, and the weather. Consequently, he emphasized the need to parametrize these subgrid-scale processes in GCMs.

Assuming that atmospheric variables can be separated into large-scale, mesoscale, and turbulent-scale, Avissar and Chen (1993) developed a set of prognostic equations applicable in large-scale atmospheric models for momentum, temperature, moisture, and any other gaseous or aerosol material, which includes both mesoscale and turbulent fluxes. They also developed prognostic equations for these mesoscale fluxes, which indicate a closure problem and, therefore, require a parametrization. For this purpose, they suggested using the mean mesoscale kinetic energy (MKE) per unit of mass. They developed a prognostic equation for MKE, and an analysis of the different terms of this equation indicated that the mesoscale vertical heat flux, the mesoscale pressure correlation, and the interaction between turbulence and mesoscale perturbations are the major terms that affect the time tendency of MKE.

The aim of the present paper is to discuss the MKE produced by mesoscale circulations that develop above deforested regions, and to demonstrate the potential use of this variable in the parametrization of subgrid-scale hydrological processes in GCMs. Thus, this study provides further insights to the basic analysis of Avissar and Chen (1993). For this purpose, in Section 2, we first summarize the set of prognostic equations for large-scale atmospheric models suggested by Avissar and Chen (1993). Then, in Section 3, we present the diurnal variation of the atmospheric planetary

boundary layer (PBL) as affected by mesoscale circulations generated by local deforestation in the Amazonian region. We also discuss qualitatively the relationship between MKE, turbulence kinetic energy (TKE), mesoscale heat fluxes, and turbulent heat fluxes. The impact of the large-scale mean wind and of the intensity of the horizontal thermal gradient on MKE, TKE, and mesoscale and turbulent heat fluxes is also investigated. Finally, in Section 4, we suggest additional studies that will need to be conducted to provide a complete parametrization of mesoscale fluxes in large-scale atmospheric models, and we discuss the new insights on the deforestation problem provided by our simulations.

The numerical simulations presented in this study were produced with the Colorado State University (CSU) Regional Atmospheric Modeling System (RAMS). The general characteristics of this mesoscale model are summarized in Pielke et al. (1992). The nonhydrostatic, anelastic option of the model was used. Turbulence is parametrized with the 2.5 level scheme suggested by Mellor and Yamada (1982). The land-surface scheme is based on the land-atmosphere interactions dynamics (LAID) originally developed by Avissar and Mahrer (1988).

2. PROGNOSTIC EQUATIONS FOR GCMs

The set of prognostic equations for large-scale atmospheric models summarized below was derived and described in detail by Avissar and Chen (1993). Thus, for further clarification, interested readers are referred to this previous study.

2.1 Prognostic equations for mean quantities

Using Einstein's summation notation, the conservation of momentum, heat, moisture, and any other gaseous or aerosol material customarily used in atmospheric models can be expressed by the following equations, respectively:

$$\frac{\partial u_i}{\partial t} + u_j \frac{\partial u_i}{\partial x_j} = -\delta_{i3}g - 2\varepsilon_{ijk}\Omega_j u_k - \frac{1}{\rho}\frac{\partial p}{\partial x_i} + \nu \frac{\partial^2 u_i}{\partial x_j^2} \quad (1)$$

$$\frac{\partial \theta}{\partial t} + u_j \frac{\partial \theta}{\partial x_j} = \nu \frac{\partial^2 \theta}{\partial x_j^2} + \psi_\theta \quad (2)$$

$$\frac{\partial q_n}{\partial t} + u_j \frac{\partial q_n}{\partial x_j} = \nu \frac{\partial^2 q_n}{\partial x_j^2} + \psi_{q_n} \quad (3)$$

and

$$\frac{\partial \chi_m}{\partial t} + u_j \frac{\partial \chi_m}{\partial x_j} = \nu \frac{\partial^2 \chi_m}{\partial x_j^2} + \psi_{\chi_m} \quad (4)$$

where u_i represents the three Cartesian wind components ($u_1 = u$ – eastward-moving; $u_2 = v$ – northward-moving; and

$u_3 = w$ – upward-moving), t is time, x_i represents the three Cartesian directions ($x_1 = x$ – eastward; $x_2 = y$ – northward; and $x_3 = z$ – upward), δ_{ij} is the Kronecker delta, g is the gravitational acceleration, ε_{ijk} is the alternating unit tensor, Ω_j represents the three Cartesian components of the angular velocity vector of the earth's rotation, ρ is the air density, p is the pressure, υ is the air kinematic viscosity, θ is the potential temperature, q_n is the specific humidity (with $n = 1,2,3$ indicating the three phases of water), χ_m is any gaseous or aerosol material (with $m = 1,2,3,\ldots,M$ indicating the various phases and chemical products of the material), and ψ_θ, ψ_{q_n}, and ψ_{χ_m} are the sources and sinks of potential temperature, specific humidity, and gaseous or aerosol material, respectively.

Let D^j be a horizontal domain ($j = 1$ – eastward; $j = 2$ – northward), which is represented in a large-scale model by a single grid element (i.e. about 300 to 500 km in current state-of-the-art GCMs), and let d^j be a horizontal domain represented by one grid element in a mesoscale model (i.e., typically 5 to 10 km). Thus, if a mesoscale model is used to simulate the domain represented by a single grid element in a GCM at a higher resolution, d^j is a subdomain of D^j, and we have:

$$D^j = \sum_{i=1}^{n} d_i^j \qquad (5)$$

with n being the number of grid points in the mesoscale model used to represent D^j.

Within a mesoscale grid element d_i^j, let ϕ be an instantaneous variable that can be partitioned into a mean, $\bar{\phi}$ and a turbulent perturbation from that mean, ϕ''. Then we have:

$$\phi = \bar{\phi} + \phi'' \qquad (6)$$

Accordingly, the conservation of momentum (eq. (1)) can be written:

$$\frac{\partial \bar{u}_i}{\partial_t} + \frac{\partial u_i''}{\partial t} + \bar{u}_j \frac{\partial \bar{u}_i}{\partial x_j} + \bar{u}_j \frac{\partial u_i''}{\partial x_j} + u_j'' \frac{\partial \bar{u}_i}{\partial x_j} + u_j'' \frac{\partial u_i''}{\partial x_j}$$
$$= -\delta_{i3}g - 2\varepsilon_{ijk}\Omega_j\bar{u}_k - 2\varepsilon_{ijk}\Omega_j u_k'' - \frac{1}{\bar{\rho} + \rho''}\frac{\partial \bar{p}}{\partial x_i}$$
$$- \frac{1}{\bar{\rho} + \rho''}\frac{\partial p''}{\partial x_i} + v\frac{\partial^2 \bar{u}_i}{\partial x_j^2} + v\frac{\partial^2 u_i''}{\partial x_j^2} \qquad (7)$$

Upon averaging this equation over d_i^j while applying Reynolds averaging rules, and using the Boussinesq approximation and the continuity equation for motion perturbations, we are left with a prognostic equation for the mean (mesoscale grid-average) wind:

$$\frac{\partial \bar{u}_i}{\partial t} + \bar{u}_j \frac{\partial \bar{u}_i}{\partial x_j} = -\delta_{i3}g - 2\varepsilon_{ijk}\Omega_j\bar{u}_k - \frac{1}{\bar{\rho}}\frac{\partial \bar{p}}{\partial x_i} + v\frac{\partial^2 \bar{u}_i}{\partial x_j^2} - \frac{\partial \overline{u_j''u_i''}}{\partial x_j} \qquad (8)$$

Assuming further that a mean variable at the mesoscale resolution, $\bar{\phi}$, consists of a mesoscale perturbation, ϕ',

superimposed on a large-scale (i.e., GCM grid-scale) mean variable, $\tilde{\phi}$, we have:

$$\bar{\phi} = \tilde{\phi} + \phi' \qquad (9)$$

and, consequently:

$$\phi = \tilde{\phi} + \phi' + \phi'' \qquad (10)$$

It is important to emphasize that ϕ' is considered as a perturbation at the scale of D^j, but is a constant in space (i.e. grid average) at the scale of d_i^j, and that $\tilde{\phi}$ is assumed constant in space at the scale of D^j (and, of course, at the scale of d_i^j).

Using this additional assumption, the prognostic equation for the mean wind (eq. (8)) can be rewritten as:

$$\frac{\partial \tilde{u}_i}{\partial t} + \frac{\partial u_i'}{\partial t} + \tilde{u}_j \frac{\partial \tilde{u}_i}{\partial x_j} + \tilde{u}_j \frac{\partial u_i'}{\partial x_j} + u_j' \frac{\partial \tilde{u}_i}{\partial x_j} + u_j' \frac{\partial u_i'}{\partial x_j}$$
$$= -\delta_{i3}g - 2\varepsilon_{ijk}\Omega_j\tilde{u}_k - 2\varepsilon_{ijk}\Omega_j u_k' - \frac{1}{\tilde{\rho} + \rho'}\frac{\partial \tilde{p}}{\partial x_i}$$
$$- \frac{1}{\tilde{\rho} + \rho'}\frac{\partial p'}{\partial x_i} + v\frac{\partial^2 \tilde{u}_i}{\partial x_j^2} + v\frac{\partial^2 u_i'}{\partial x_j^2} - \frac{\partial \tilde{S}_{ij}}{\partial x_j} - \frac{\partial S_{ij}'}{\partial x_j} \qquad (11)$$

where the turbulent momentum fluxes are written $\bar{S}_{ij} = \overline{u_i''u_j''}$.

Averaging the above equation over D^j while applying Reynolds averaging rules, and using the Boussinesq approximation and the continuity equation for motion perturbations, we can define the following prognostic equation for the large-scale mean wind:

$$\frac{\partial \tilde{u}_i}{\partial t} + \tilde{u}_j \frac{\partial \tilde{u}_i}{\partial x_j} = -\delta_{i3}g - 2\varepsilon_{ijk}\Omega_j\tilde{u}_k - \frac{1}{\tilde{\rho}}\frac{\partial \tilde{p}}{\partial x_i} + v\frac{\partial^2 \tilde{u}_i}{\partial x_j^2} - \frac{\partial \tilde{S}_{ij}}{\partial x_j} - \frac{\partial \langle u_j'u_i' \rangle}{\partial x_j} \qquad (12)$$

where the $\langle \; \rangle$ symbol is the averaging operator over D^j (noting that $\langle \tilde{\phi} \rangle = \tilde{\phi}$, and that $\langle \phi' \rangle = 0$).

Adopting a similar procedure, we developed the following prognostic equations for the other synoptic-scale mean atmospheric variables:

$$\frac{\partial \tilde{\theta}}{\partial t} + \tilde{u}_j \frac{\partial \tilde{\theta}}{\partial x_j} = v\frac{\partial^2 \tilde{\theta}}{\partial x_j^2} + \tilde{\psi}_\theta - \frac{\partial \tilde{S}_{\theta j}}{\partial x_j} - \frac{\partial \langle u_j'\theta' \rangle}{\partial x_j} \qquad (13)$$

$$\frac{\partial \tilde{q}}{\partial t} + \tilde{u}_j \frac{\partial \tilde{q}}{\partial x_j} = v\frac{\partial^2 \tilde{q}}{\partial x_j^2} + \tilde{\psi}_q - \frac{\partial \tilde{S}_{qj}}{\partial x_j} - \frac{\partial \langle u_j'q' \rangle}{\partial x_j} \qquad (14)$$

$$\frac{\partial \tilde{\chi}}{\partial t} + \tilde{u}_j \frac{\partial \tilde{\chi}}{\partial x_j} = v\frac{\partial^2 \tilde{\chi}}{\partial x_j^2} + \tilde{\psi}_\chi - \frac{\partial \tilde{S}_{\chi j}}{\partial x_j} - \frac{\partial \langle u_j'\chi' \rangle}{\partial x_j} \qquad (15)$$

where the subscripts n in q_n and m in χ_m have been dropped, for simplicity.

It is important to mention that the above set of equations (i.e. eqs. (12)–(15)) should be used in GCMs (and other large-scale atmospheric models) to prognosticate mean atmospheric variables. It clearly indicates the contribution of turbulence and mesoscale perturbations to the mean variables. While current state-of-the-art GCMs include a para-metrization of turbulent fluxes (yet incorrectly assuming horizontal homogeneity) none of them account for the mesoscale fluxes represented by the last term in each of these equations. As emphasized by Pielke et al. (1991) and Avissar and Chen (1993), however, these mesoscale fluxes are often even more important than the turbulent fluxes, and should not be omitted. But clearly, this procedure introduces another set of unknowns to the basic set of equations, which obviously requires a solution. For this purpose, we also developed prognostic equations for these mesoscale fluxes.

2.2 Prognostic equations for mesoscale fluxes

Subtracting eq. (12) from eq. (11) (after applying Boussinesq approximation and a linearized form of the ideal gas law, i.e. $\rho'/\bar{\rho} = -\theta'_v/\bar{\theta}_v$ with θ_v being the virtual potential temperature) the following prognostic equation for the mesoscale wind perturbation is obtained:

$$
\frac{\partial u'_i}{\partial t} + \tilde{u}_j \frac{\partial u'_i}{\partial x_j} + u'_j \frac{\partial \tilde{u}_i}{\partial x_j} + u'_j \frac{\partial u'_i}{\partial x_j}
$$
$$
= \delta_{i3} \frac{\theta'_v}{\bar{\theta}_v} g - 2\varepsilon_{ijk}\Omega_j u'_k - \frac{1}{\bar{\rho}} \frac{\partial p'}{\partial x_i} + v \frac{\partial^2 u'_i}{\partial x_j^2} - \frac{\partial S'_{ij}}{\partial x_j} + \frac{\partial \langle u'_j u'_i \rangle}{\partial x_j} \quad (16)
$$

Multiplying all terms of this equation by u'_m, we obtain:

$$
u'_m \frac{\partial u'_i}{\partial t} + u'_m \tilde{u}_j \frac{\partial u'_i}{\partial x_j} + u'_m u'_j \frac{\partial \tilde{u}_i}{\partial x_j} = u'_m u'_j \frac{\partial u'_i}{\partial x_j}
$$
$$
= \delta_{i3} \frac{g}{\bar{\theta}_v} u'_m \theta'_v - 2\varepsilon_{ijk}\Omega_j u'_m u'_k - \frac{u'_m}{\bar{\rho}} \frac{\partial p'}{\partial x_i}
$$
$$
+ u'_m v \frac{\partial^2 u'_i}{\partial x_j^2} - u'_m \frac{\partial S'_{ij}}{\partial x_j} + u'_m \frac{\partial \langle u'_j u'_i \rangle}{\partial x_j} \quad (17)
$$

An identical equation to eq. (17) can be written by interchanging the i and m indices. Adding finally these two identical equations using the product rule and averaging over D^j, the following prognostic equation for mesoscale momentum fluxes is derived:

$$
\frac{\partial \langle u'_i u'_m \rangle}{\partial t} + \tilde{u}_j \frac{\partial \langle u'_i u'_m \rangle}{\partial x_j} + \langle u'_m u'_j \rangle \frac{\partial \tilde{u}_i}{\partial x_j} + \langle u'_i u'_j \rangle \frac{\partial \tilde{u}_m}{\partial x_j} + \frac{\partial \langle u'_i u'_j u'_m \rangle}{\partial x_j}
$$
$$
= + \frac{g}{\bar{\theta}_v} \left(\delta_{i3} \langle u'_m \theta'_v \rangle + \delta_{m3} \langle u'_i \theta'_v \rangle \right) - \left\langle \frac{u'_m}{\bar{\rho}} \frac{\partial p'}{\partial x_i} \right\rangle - \left\langle \frac{u'_i}{\bar{\rho}} \frac{\partial p'}{\partial x_m} \right\rangle
$$
$$
+ \left\langle u'_m v \frac{\partial^2 u'_i}{\partial x_j^2} \right\rangle + \left\langle u'_i v \frac{\partial^2 u'_m}{\partial x_j^2} \right\rangle - \left\langle u'_m \frac{\partial S'_{ij}}{\partial x_j} \right\rangle - \left\langle u'_i \frac{\partial S'_{mj}}{\partial x_j} \right\rangle \quad (18)
$$

A similar procedure was adopted to produce prognostic equations for the mesoscale perturbation of potential temperature, specific humidity, and gaseous or aerosol materials. Multiplying each of these equations by the mesoscale wind perturbation (u'_i) and adding the resulting equation to eq. (16), which is itself multiplied by the corresponding meso-scale perturbation of potential temperature, specific humid-

ity, and gaseous or aerosol materials, we developed the following prognostic equations for the mesoscale fluxes of potential temperature, specific humidity, and gaseous or aerosol materials:

$$
\frac{\partial \langle u'_i \theta' \rangle}{\partial t} + \tilde{u}_j \frac{\partial \langle u'_i \theta' \rangle}{\partial x_j} + \langle u'_j \theta' \rangle \frac{\partial \tilde{u}_i}{\partial x_j} + \langle u'_i u'_j \rangle \frac{\partial \tilde{\theta}}{\partial x_j} + \frac{\partial \langle u'_i u'_j \theta' \rangle}{\partial x_j}
$$
$$
= + \delta_{i3} \frac{g}{\bar{\theta}_v} \langle \theta' \theta'_v \rangle - 2\varepsilon_{ijk}\Omega_j \langle u'_k \theta' \rangle - \left\langle \frac{\theta'}{\bar{\rho}} \frac{\partial p'}{\partial x_i} \right\rangle
$$
$$
+ \langle u'_i \psi'_\theta \rangle + \left\langle \theta' v \frac{\partial^2 u'_i}{\partial x_j^2} \right\rangle + \left\langle u'_i v \frac{\partial^2 \theta'}{\partial x_j^2} \right\rangle
$$
$$
- \left\langle \theta' \frac{\partial S'_{ij}}{\partial x_j} \right\rangle - \left\langle u'_i \frac{\partial S'_{\theta j}}{\partial x_j} \right\rangle \quad (19)
$$

$$
\frac{\partial \langle u'_i q' \rangle}{\partial t} + \tilde{u}_j \frac{\partial \langle u'_i q' \rangle}{\partial x_j} + \langle u'_j q' \rangle \frac{\partial \tilde{u}_i}{\partial x_j} + \langle u'_i u'_j \rangle \frac{\partial \tilde{q}}{\partial x_j} + \frac{\partial \langle u'_i u'_j q' \rangle}{\partial x_j}
$$
$$
= + \delta_{i3} \frac{g}{\bar{\theta}_v} \langle q' \theta'_v \rangle - 2\varepsilon_{ijk}\Omega_j \langle u'_k q' \rangle - \left\langle \frac{q'}{\bar{\rho}} \frac{\partial p'}{\partial x_i} \right\rangle
$$
$$
+ \langle u'_i \psi'_q \rangle + \left\langle q' v \frac{\partial^2 u'_i}{\partial x_j^2} \right\rangle + \left\langle u'_i v \frac{\partial^2 q'}{\partial x_j^2} \right\rangle
$$
$$
- \left\langle q' \frac{\partial S'_{ij}}{\partial x_j} \right\rangle - \left\langle u'_i \frac{\partial S'_{qj}}{\partial x_j} \right\rangle \quad (20)
$$

$$
\frac{\partial \langle u'_i \chi' \rangle}{\partial t} + \tilde{u}_j \frac{\partial \langle u'_i \chi' \rangle}{\partial x_j} + \langle u'_j \chi' \rangle \frac{\partial \tilde{u}_i}{\partial x_j} + \langle u'_i u'_j \rangle \frac{\partial \tilde{\chi}}{\partial x_j} + \frac{\partial \langle u'_i u'_j \chi' \rangle}{\partial x_j}
$$
$$
= + \delta_{i3} \frac{g}{\bar{\theta}_v} \langle \chi' \theta'_v \rangle - 2\varepsilon_{ijk}\Omega_j \langle u'_k \chi' \rangle - \left\langle \frac{\chi'}{\bar{\rho}} \frac{\partial p'}{\partial x_i} \right\rangle
$$
$$
+ \langle u'_i \psi'_\chi \rangle + \left\langle \chi' v \frac{\partial^2 u'_i}{\partial x_j^2} \right\rangle + \left\langle u'_i v \frac{\partial^2 \chi'}{\partial x_j^2} \right\rangle
$$
$$
- \left\langle \chi' \frac{\partial S'_{ij}}{\partial x_j} \right\rangle - \left\langle u'_i \frac{\partial S'_{\chi j}}{\partial x_j} \right\rangle \quad (21)
$$

The set of equations given by eqs. (12)–(15) and (18)–(21) is obviously not closed. For instance, the triple correlation terms $\langle u'_i u'_j u'_m \rangle$, $\langle u'_i u'_j \theta' \rangle$, $\langle u'_i u'_j q' \rangle$ and $\langle u'_i u'_j \chi' \rangle$, which appear in eqs. (18), (19), (20), and (21) respectively, are new unknowns that need to be solved. Prognostic equations can be developed for these unknowns, which generate quadruple correlation terms, and so on. This well-known problem in turbulence, which is referred to as the 'closure problem', needs to be addressed using appropriate parametrizations. Avissar and Chen (1993) suggested the use of MKE for such a parametrization, and developed a prognostic equation for this variable.

2.3 Prognostic equations for mesoscale kinetic energy (MKE)

Multiplying all the terms of eq. (16) by u'_i, averaging over D^j while applying Reynolds averaging rules, and using the continuity equation for perturbation motions, we obtain the following prognostic equation for the mean MKE per unit mass ($\tilde{E}$):

$$\underbrace{\frac{\partial \tilde{E}}{\partial t}}_{I} + \underbrace{\tilde{u}_j \frac{\partial \tilde{E}}{\partial x_j}}_{II} = \underbrace{\frac{g}{\bar{\theta}_v} \langle w'\theta_v' \rangle}_{III} - \underbrace{\langle u_j'u_i' \rangle \frac{\partial \tilde{u}_i}{\partial x_j}}_{IV} - \underbrace{\frac{\partial \langle u_j'E \rangle}{\partial x_j}}_{V} - \underbrace{\frac{1}{\bar{\rho}} \frac{\partial \langle u_i'p' \rangle}{\partial x_i}}_{VI}$$

$$+ \underbrace{\left\langle u_i' v \frac{\partial^2 u_i'}{\partial x_j^2} \right\rangle}_{VII} - \underbrace{\left\langle u_i' \frac{\partial S_{ij}'}{\partial x_j} \right\rangle}_{VIII} \qquad (22)$$

where

$$\tilde{E} = 0.5 \langle u_i'^2 \rangle \qquad (23)$$

and Term I is the time tendency of $\tilde{E}$; Term II is the advection by the large-scale mean wind; Term III is the temperature flux term; Term IV is the shear created by the gradients of the large-scale mean wind; Term V is the transport of $\tilde{E}$ by mesoscale wind perturbations; Term VI is the pressure correlation term; Term VII is the viscous dissipation of MKE; and Term VIII is the turbulence correlation term.

Avissar and Chen (1993) discussed the relative importance of these terms, and concluded that Term III, Term VI, and Term VIII are mostly affecting the time tendency of MKE.

3. SIMULATIONS OF LOCAL DEFORESTATION IN AMAZONIA

Deforestation is occurring at an alarming pace in tropical regions and is also frequent in mid-latitudes. Over the past years, much effort has been made to evaluate the various aspects and consequences of this anthropological activity (Dickinson, 1987). In particular, several simulations with GCMs have been produced to investigate climatological aspects of tropical rainforest deforestation (e.g. Dickinson and Henderson-Sellers, 1988; Henderson-Sellers et al., 1988; Shukla et al., 1990). In general, these simulations predict a net warming, a reduction of precipitation, and an extension of the dry season in this region. As a result, serious ecological implications are expected.

Yet it is important to realize that due to the resolvable scale of the GCMs, these simulations must assume a deforestation of the entire Amazonian basin. In reality, however, the process of deforestation is progressive, producing patches of deforested areas, as illustrated in the satellite map of this region, shown in Fig. 1. To emphasize the importance of mesoscale fluxes and their relation to MKE with a practical example, RAMS was used to simulate this type of patchy deforestation in this region.

3.1 Simulations set-up

The simulated domain was 250 km wide, and was located at latitude 0°. This domain was represented in the model by 50 grid elements with a horizontal grid resolution of 5 km. The deforested area, which was 50 km wide, was located in the middle of the domain. Thus, it was flanked by 100 km of forests. The atmosphere was simulated up to a height of 10 km (i.e. roughly the tropopause), and was represented by 18 grid elements with a high grid resolution near the ground surface, and progressively less resolution with height in the atmosphere.

We assumed an initial soil temperature of 300 K, a soil water content at field capacity in the forest area, and a soil water content of 4% of the soil water content at field capacity in the deforested area. A potential temperature lapse of 3.5 K/km, a relative humidity of 95% at all heights in the atmosphere, and a southerly background wind of 0.5 m/s were also used for the initialization of the model. Furthermore, the forest was assumed to consist of a very dense canopy, which absorbed all the solar radiation not reflected by the canopy (i.e. zero transmissivity to the soil surface).

The numerical integration for the simulations was started at 0600 a.m. (local standard time), which corresponds to the time when the sensible heat flux becomes effective in the development of the convective PBL on sunny summer days. The time step of the numerical integration was 15 s.

3.2 Impact of deforestation on the PBL

Fig. 2 presents 2-D sections of the meteorological fields (u – the west-east horizontal component of the wind parallel to the domain, w – the vertical component of the wind, θ – the potential temperature, and r – the mixing ratio) obtained at 9 a.m., 3 p.m., 9 p.m., and 3 a.m. next morning. The part of the simulated domain covered by dense forest is indicated in these sections by a black underbar, and only the lower 3.5 km of the atmosphere is displayed.

The regional circulations depicted by the horizontal and vertical components of the wind result from the differential heating produced by the two very distinct surfaces subjected to the same solar radiation input. In the deforested land (assumed bare and relatively dry), most of the radiative energy received from the sun and the atmosphere is used to heat the atmosphere and the ground. However, in the forest area (assumed unstressed), a large part of this radiative energy is used for evapotranspiration (e.g. Avissar, 1992). The faster heating rate above the deforested land surface generates a vigorous turbulent mixing and an unstable, stratified PBL, as can be seen from the sections of potential temperature in Fig. 2. On the contrary, the slower heating rate above the transpiring forest limits the development of

0 10   20   30 km

Fig. 1 Satellite image of local deforestation in the Brazil Amazonian Tropical Forest (by courtesy of C.A. Nobre, Center for Weather Prediction and Climate Studies, Brazil).

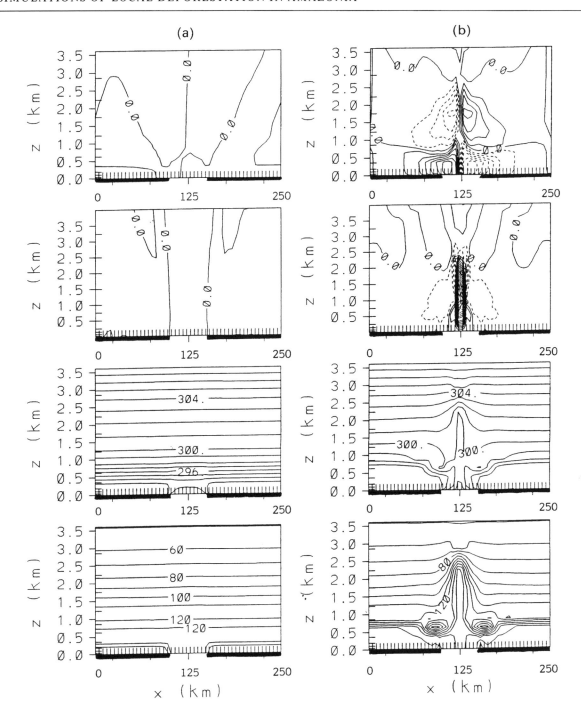

Fig. 2 Two-dimensional sections of (i) the horizontal wind component parallel to the domain (*u*) in m/s, positive eastward; (ii) the vertical wind component (*w*) in cm/s, positive upward; (iii) the potential temperature (θ) in K; and (iv) the mixing ratio ($r \times 10^4$), obtained in the planetary boundary layer that develops over a deforested area of the Amazonian region, during mid-summer time at (a) 9 a.m., (b) 3 p.m., (c) 9 p.m., and (d) 3 a.m. Forest is indicated by a dark underbar. Background wind is 0.5 m/s blowing from south. Solid lines indicate positive values, and dashed lines indicate negative values.

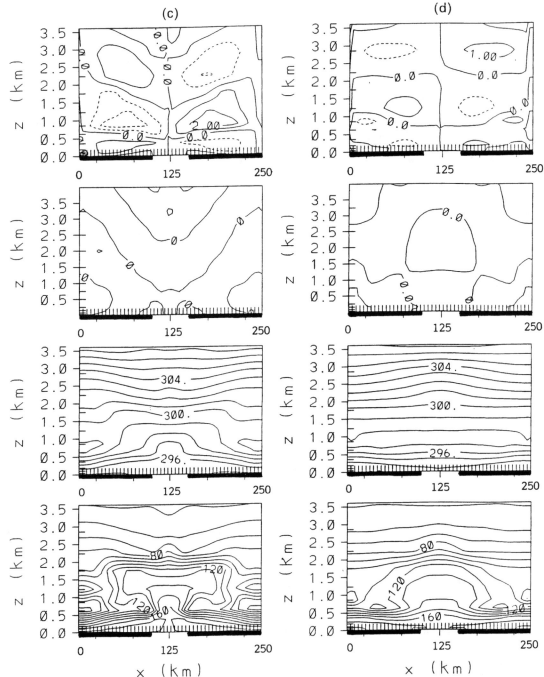

Fig. 2 *(cont).*

the PBL, which remains shallow above the forest. This creates a pressure gradient between the two areas, which generates the circulations from the relatively cold to the relatively warm area.

Because the differential heating is closely related to the redistribution of solar energy absorbed at the ground surface, a diurnal pattern is clearly evident in Fig. 2. In the early morning (9 a.m.), the amount of solar radiation reaching the land surface is relatively small and, as a result, only a weak thermal and pressure gradient is created between the forest and the deforested areas. These gradients increase gradually

during the morning hours and dissipate during the afternoon and night hours. The time-lag of the maximum intensity of the mesoscale circulation that develops over the domain is mainly due to the time response of the PBL to the heating and cooling, which occur mostly at the ground surface.

The transpiration from the forest provides a supply of moisture which significantly increases the amount of water in the shallow PBL, as can be seen from the 2-D sections of the mixing ratio presented in Fig. 2. This moisture is advected by the generated mesoscale flow which strongly converges toward the deforested area, where it is well mixed within the relatively deep convective boundary layer. This process may eventually generate convective clouds and precipitation under appropriate synoptic-scale atmospheric conditions, as already emphasized in previous studies (Avissar, 1991) and is discussed in more detail later in this paper.

3.3 Mesoscale kinetic energy (MKE) versus turbulence kinetic energy (TKE)

Fig. 3 shows the 2-D sections of MKE and TKE produced in this deforestation simulation, at the same time periods used to illustrate the atmospheric variables in Fig. 2 (i.e. 9 a.m., 3 p.m., 9 p.m., and 3 a.m.). Noting that the same units and contour intervals have been adopted for MKE and TKE in these 2-D sections, one can clearly appreciate the importance of MKE as compared to TKE. While the TKE maximum at 3 p.m. is about 4 m^2/s^2, the corresponding MKE maximum (which is approximately the daily maximum) is about 23 m^2/s^2 for this particular simulation.

It is important to emphasize that RAMS calculates meso-scale perturbations relative to reference atmospheric conditions, which are assumed to represent the large-scale conditions, and that are maintained constant during the simulations. Since these large-scale conditions are not updated at each time step of the numerical integration to account for the contribution of the mesoscale flux divergences, the mesoscale perturbations shown in our figures are not exactly the same mesoscale perturbations as defined in our equations, which should average to zero when integrated over D^j. Thus, this bias introduces some distortion of the various atmospheric fields illustrated in our figures. For example, updating the mean wind in the simulation discussed here would cause a reduction of the mesoscale wind perturbations and, consequently, a reduction of the MKE. This aspect, however, does not affect significantly the general trend of our results, and is discussed in more detail in Section 3.5.

It is interesting to note that TKE is distributed more or less uniformly over the forest and deforested parts of the domain, showing a relatively strong turbulence activity over the deforested (dry) land (which is fed by strong surface sensible heat flux and buoyancy), but a limited activity over the forest. The maximum value of TKE is found at approximately 1/3 the height of the PBL, as expected from previous observational and modeling studies (e.g. Deardorff, 1974; André et al., 1978; Yamada and Mellor, 1975; Therry and Lacarrère, 1983).

On the other hand, MKE is found only in that part of the domain which is affected by the generated mesoscale circulation. Thus, during the afternoon hours, it is concentrated within a domain of about 100 km around the deforested land, and presents two zones of large intensity. One is in the lower part of the PBL and corresponds to the flow from the forest, which is relatively cold, to the deforested area, which is relatively warm. The second is weaker but deeper, and corresponds to the return flow that develops in the upper part of the PBL. Later in the day, the mesoscale flow gradually expands over a larger part of the domain, and dissipates strongly, due mostly to the dissipation of the thermal gradient in the cooling atmosphere (see Fig. 2). Correspondingly, the MKE spreads over a larger domain, and strongly dissipates.

From a large-scale modeling perspective, the importance of MKE is perhaps better illustrated in Fig. 4, which depicts the profiles of horizontal averages (over D^j) of MKE and TKE, again at 9 a.m., 3 p.m., 9 p.m., and 3 a.m. The MKE maximum at 3 p.m. is about 4 times the TKE maximum, and penetrates much higher in the atmosphere than the TKE (about 3.5 km as compared to about 0.75 km). Furthermore, the time residency of MKE in the PBL is also much longer than that of TKE, which dissipates rapidly during the later afternoon and the evening. Note that the shape of the TKE profiles simulated at the different hours in this numerical experiment is quite similar to TKE profiles simulated and observed in other investigations (e.g. Therry and Lacarrère, 1983).

3.4 Mesoscale versus turbulent heat fluxes

Fig. 5 shows the vertical profiles of mean (i.e. horizontally averaged) mesoscale and turbulent kinematic sensible and latent heat fluxes obtained in the above described simulation. The mesoscale vertical sensible heat flux is the covariance between the vertical component of the mesoscale circulation (w') and the perturbation of potential temperature (θ'). As one can see in Fig. 2, in general, θ' is positive in that region of the PBL where a strong, positive w' has developed (i.e. the lower 2 km of the atmosphere). This results in a positive product $w'\theta'$, indicating a net transport of heat upward. However, in the upper part of the PBL (where the return flow is dominant), the sign of w' and θ' is usually opposite. This results in a negative mesoscale sensible heat flux, which corresponds to a net transport of heat downward. The general shape of the profile of turbulent sensible heat flux is similar to that observed, simulated, and discussed in detail by many others (e.g. Deardorff, 1974; André et al., 1978; Stull, 1988).

The profile of mesoscale vertical latent heat flux is very different in the lower part and the upper part of the PBL. In general, in the lower part of the PBL, positive perturbations of mixing ratio are found above the wet land, where w' is typically negative. In this region, however, turbulent fluxes are positive, indicating a net turbulent transport of moisture

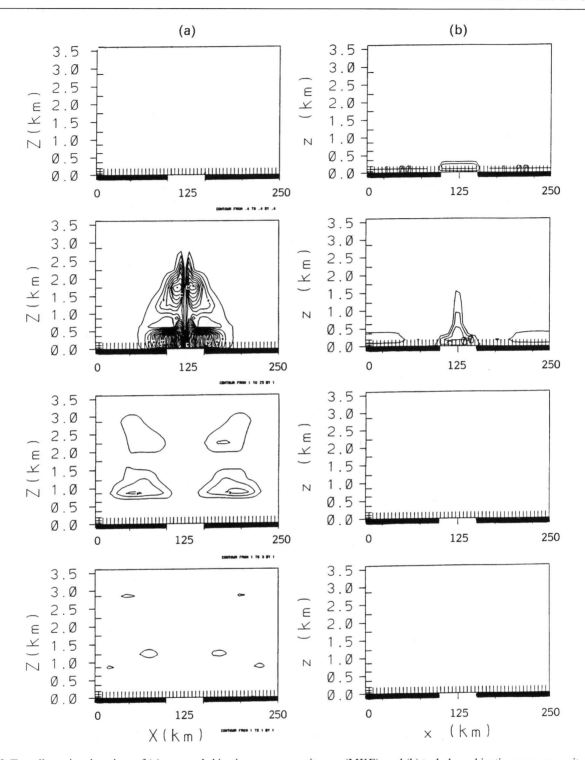

Fig. 3 Two-dimensional sections of (a) mesoscale kinetic energy per unit mass (MKE), and (b) turbulence kinetic energy per unit mass (TKE) obtained in the planetary boundary layer that develops over a deforested area of the Amazonian region, during mid-summer time at (i) 9 a.m., (ii) 3 p.m., (iii) 9 p.m., and (iv) 3 a.m. Units are m²/s². Forest is indicated by a dark underbar. Background wind is 0.5 m/s blowing from south.

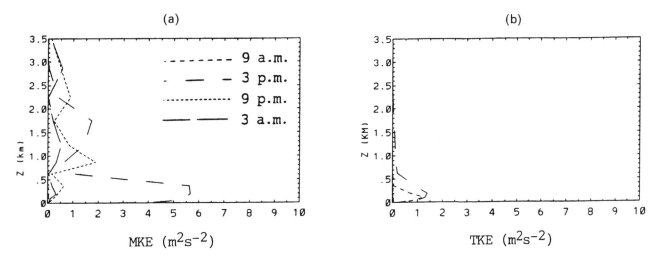

Fig. 4 Profiles of (a) horizontal mean mesoscale kinetic energy per unit mass, and (b) horizontal mean turbulence kinetic energy per unit mass obtained in the planetary boundary layer that develops over a deforested area of the Amazonian region, during mid-summer time at 9 a.m., 3 p.m., 9 p.m., and 3 a.m. Background wind is 0.5 m/s blowing from south.

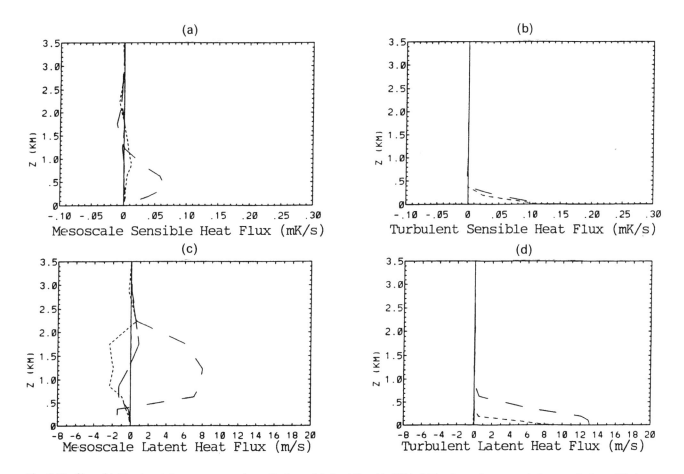

Fig. 5 Profiles of (a) horizontal mean mesoscale vertical sensible heat flux ($\langle w'\theta' \rangle$), (b) horizontal mean turbulent vertical sensible heat flux ($\langle \overline{w''\theta''} \rangle$), (c) horizontal mean mesoscale vertical latent heat flux ($\langle w'q' \rangle$), and (d) horizontal mean turbulent vertical latent heat flux ($\langle \overline{w''\theta''} \rangle$), obtained in the planetary boundary layer that develops over a deforested area of the Amazonian region, during mid-summer time at 9 a.m., 3 p.m., 9 p.m., and 3 a.m. (legend is as in Fig. 4). Background wind is 0.5 m/s blowing from south.

upward. Thus, the mesoscale vertical flow tends to act as a counter-gradient term in this part of the PBL.

Above the dry land, where perturbations of mixing ratio are negative, w' is mostly positive. Thus, in this part of the PBL, the covariance $\langle w'r' \rangle$ is also negative, indicating a mesoscale transport of relatively dry air upward. It is important to remember that the moisture evaporated from the wet land is transported toward the dry land (i.e. the covariance $\langle u'r' \rangle$ is positive and very strong), and even though the moisture above the dry land is smaller than above the wet land (i.e. a negative perturbation of mixing ratio), this moisture is still larger than the moisture at higher elevation in this part of the domain. As a result, the negative covariance does not necessarily indicate a flux of moisture downward.

The moisture transported upward in the PBL by the combined effect of turbulence and mesoscale vertical flow, creates a positive mesoscale perturbation above a height of approximately 400 m. As a result, moisture is drained from the lower part of the PBL and dampens the upper part of the PBL.

Among the different fluxes produced by the mesoscale circulation that develops as a result of a landscape discontinuity, the moisture flux is probably the most valuable for parametrization purposes in large-scale models. This is because the dampening of the PBL is likely to form clouds, which affect the radiation regime, precipitation, and, as a result, the hydrologic cycle. For instance, in a recent study, Bougeault et al. (1991) simulated with the French Weather Service limited-area numerical model (PERIDOT) the formation of clouds next to the boundary between a forest and a crop area. Their results were supported by satellite images and observations of various meteorological parameters in the PBL.

A qualitative relationship between the vertical distribution of MKE and the profile of mesoscale vertical latent heat flux is evident when comparing Fig. 4 and Fig. 5. At the various hours of the day, the maximum MKE in the lower part of the PBL clearly corresponds to the maximum negative latent heat flux in the PBL, while the maximum MKE in the upper part of the PBL clearly corresponds to the maximum positive latent heat flux in the PBL.

It is interesting to note that turbulence and mesoscale flow complete each other remarkably well to transport heat and moisture upward in the PBL. As clearly depicted in Fig. 5, turbulence is mostly effective close to the ground surface and in the lower part of the PBL, while the contribution of the mesoscale flow is mostly effective higher in the PBL. Integrating with height the profiles of mesoscale and turbulent heat fluxes shown in Fig. 5, one can see that the mesoscale fluxes are about two times stronger than the turbulent fluxes, emphasizing again the importance of mesoscale fluxes as compared to turbulent fluxes.

3.5 Mean wind impact

In order to evaluate the impact of background-wind intensity on MKE and the mesoscale fluxes, we used RAMS to produce another simulation, assuming the same conditions as in the previous simulation, but for the background wind. In this case, the model was initialized with an easterly wind of $u = 5$ m/s.

Fig. 6 presents 2-D sections of the different atmospheric fields, at 9 a.m., 3 p.m., 9 p.m., and 3 a.m. next morning. As for the zero-wind case illustrated in Fig. 2, a diurnal variation of these fields is obtained.

As one can see, a background wind in the direction of the mesoscale flow tends to reduce the thermal gradient generated between the wet forest and the deforested, dry region of the domain, thus reducing the intensity of the mesoscale circulation. On the other hand, a background wind opposite to the development of the mesoscale flow increases the thermal gradient and, as a result, intensifies the mesoscale circulation. This creates a deformation of the atmospheric fields, which is reflected in the 2-D sections of MKE and TKE depicted in Fig. 7. Interestingly, this process has only a limited impact on the vertical profiles of horizontal averages of MKE and TKE presented in Fig. 8, as well as on the mesoscale and turbulent fluxes shown in Fig. 9.

It must be noted that because of the symmetry of the simulated domain, initializing the model with a westerly wind of $u = -5$ m/s produces similar effects, but with a deformation of the different atmospheric fields oriented westward (not shown here, for brevity). Furthermore, initializing the model with southerly or northerly winds has almost no impact at all on the various atmospheric fields and fluxes. Stronger winds were also used to initialize the model. Up to 15 m/s, the deformation of the atmospheric fields produced by the wind was similar, but proportionally stronger than that produced by a wind of 5 m/s.

Thus, from these simulations, it seems that the impact of the background wind is not significant, and could be ignored for parametrization purposes. Yet Avissar and Chen (1993), who investigated nonsymmetric landscape discontinuities, found a relatively important background wind effect on MKE and mesoscale fluxes (as expected from our results, when considering the impact of the background wind on each of the two mesoscale circulations that develop over the deforested region). Further analysis is still required to clarify this complex relation.

3.6 Horizontal thermal gradient impact

As discussed above, the mesoscale circulations produced in our simulations develop as a result of different redistributions of solar energy on dry (deforested) and wet (forested)

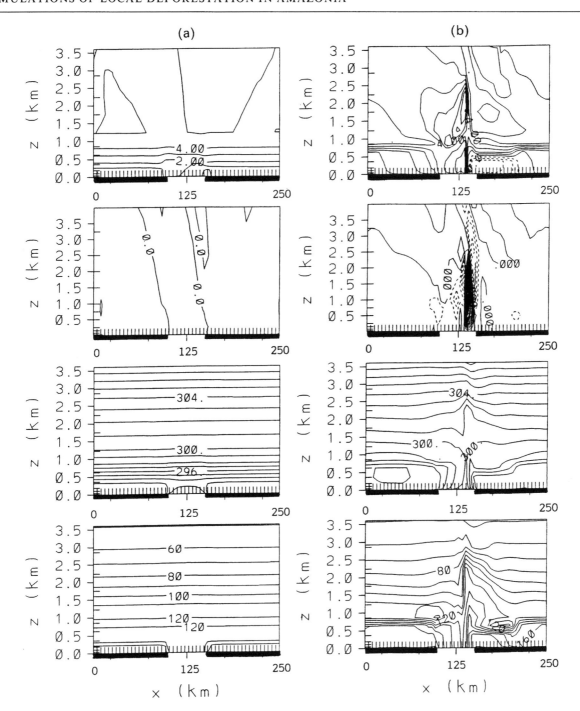

Fig. 6 Same as Fig. 2, but for an easterly background wind of 5 m/s.

land. Segal et al. (1988), Avissar and Pielke (1989), and Avissar (1991), showed that the intensity of this type of circulation is related (nonlinearly) to the moisture contrast at the landscape discontinuity, which results in a horizontal thermal gradient in the atmosphere above the ground surface. The stronger the contrast, the stronger the mesoscale flow.

To demonstrate the impact of this moisture/thermal gradient on MKE and mesoscale fluxes, RAMS was used to produce two additional simulations. The same initial conditions as in the previous simulations were used again, except for the soil moisture in the deforested area. In one case, soil moisture was increased to only 25% of the soil water content at field capacity. In the other case, soil moisture was increased to 50% of the soil water content at field capacity.

Fig. 10 presents 2-D sections of the different atmospheric fields, at 3 p.m, for these two additional simulations. As for the previous simulations, a diurnal pattern of these fields is

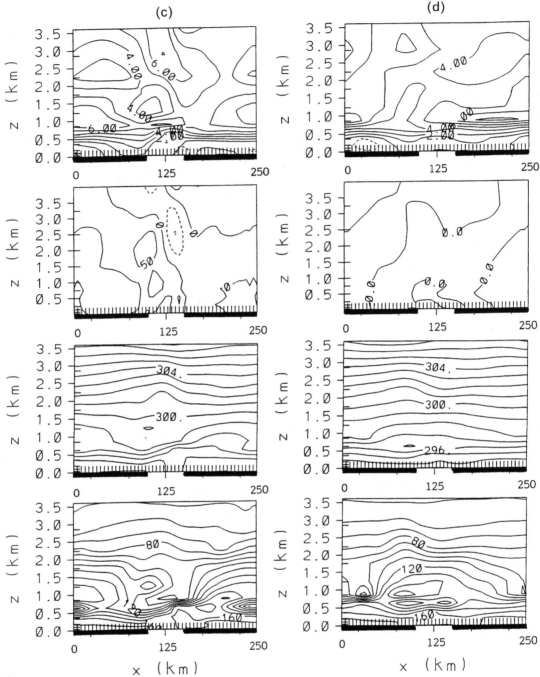

Fig. 6 (cont).

obtained, but is not shown for brevity. Clearly, the non-linearity of the relationship between the moisture gradient and the intensity of the mesoscale circulation is well emphasized. Indeed, while the circulation is eliminated when the moisture gradient is zero (i.e. when the deforested area has been watered by precipitation), even for a moisture content of only 50% of the soil water content at field capacity in the dry land, no mesoscale effect is apparent. This is because there is still enough water available at the ground surface for evaporation. However, when the soil moisture is increased to just 25% of the soil water content at field capacity in the

deforested area, the horizontal and vertical components of the mesoscale circulation are already relatively strong.

This effect is reproduced in the various 2-D sections and vertical profiles of horizontal averages shown in Figs. 11–14. When there is no mesoscale flow, there is no MKE, and there are no mesoscale heat fluxes. When a limited mesoscale flow develops, however, MKE and mesoscale fluxes develop proportionally, emphasizing again the ability of MKE to bridge between the landscape discontinuity, the mesoscale flow, and the mesoscale fluxes.

It is also interesting to look at the profiles of TKE and

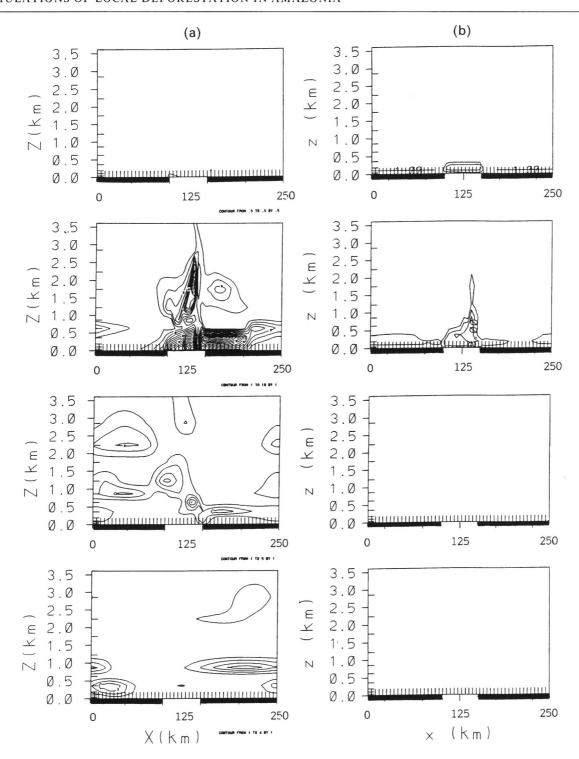

Fig. 7 Same as Fig. 3, but for an easterly background wind of 5 m/s.

turbulent heat fluxes, produced by these additional simulations. In particular, when the entire domain is wet enough to preclude the formation of a mesoscale circulation, moisture and heat are transported upward in the PBL by turbulence only. Thus, the profiles of mean turbulent heat fluxes shown in Fig. 14 represent the maximum fluxes possible at this particular location, for this time of the year, and these

atmospheric background conditions, when there is no mesoscale activity in the region. Comparing these profiles with those shown in Fig. 5, one can clearly see the significant impact that the mesoscale circulations have on the vertical distribution of turbulent fluxes, in addition to their impact on the mesoscale heat fluxes. Integrating with height the profiles of turbulent latent heat flux shown in Fig. 5 and Fig. 14, we

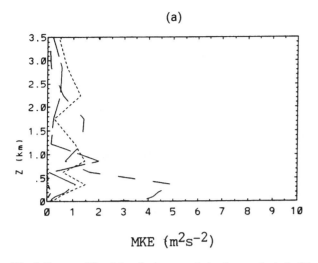

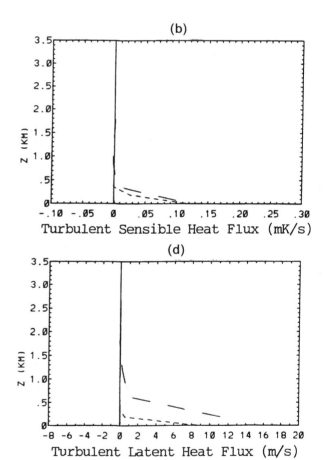

Fig. 8 Same as Fig. 4, but for an easterly background wind of 5 m/s.

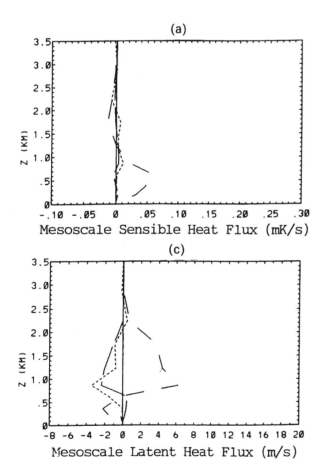

Fig. 9 Same as Fig. 5, but for an easterly background wind of 5 m/s.

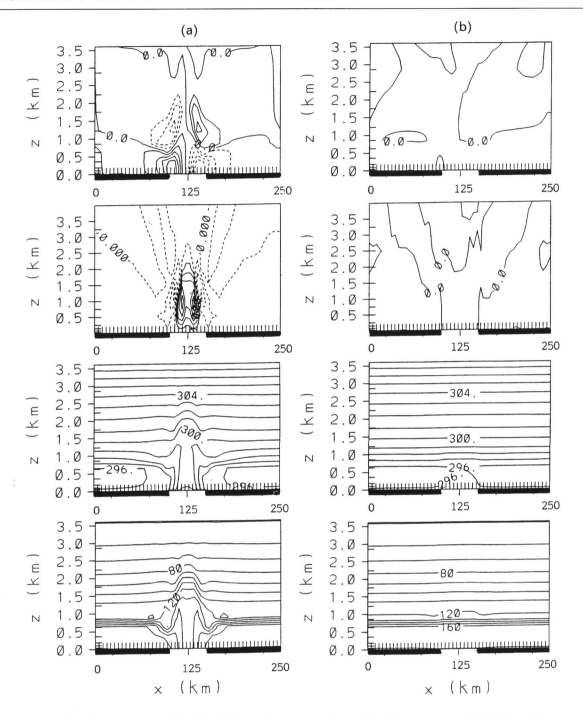

Fig. 10 Same as Fig. 2, but at 3 p.m. only for (a) a deforested area watered to a soil moisture content of 25% of the soil water content at field capacity, and (b) a deforested area watered to a soil moisture content of 50% of the soil water content at field capacity.

find that with the development of a mesoscale circulation, the integrated turbulent latent heat flux is about 10–20% stronger than over the homogeneous wet domain, even though evaporation only occurs over 200 km of the 250 km of the domain (i.e. 80%)!

Thus, these numerical simulations suggest that the moisture/thermal gradient is an important parameter that will need to be considered for parametrization purposes. Satellite observations are likely to provide helpful information to identify such gradients at the global scale (e.g. Avissar, 1992). Furthermore, these numerical experiments illustrate the significant impact of the interaction between turbulence and mesoscale perturbations on mesoscale fluxes as well as on turbulent fluxes. Therefore, this clearly emphasizes that current state-of-the-art large-scale models, which ignore the impact of landscape (subgrid-scale) heterogeneity on mesoscale fluxes, are likely to represent turbulent fluxes poorly as well.

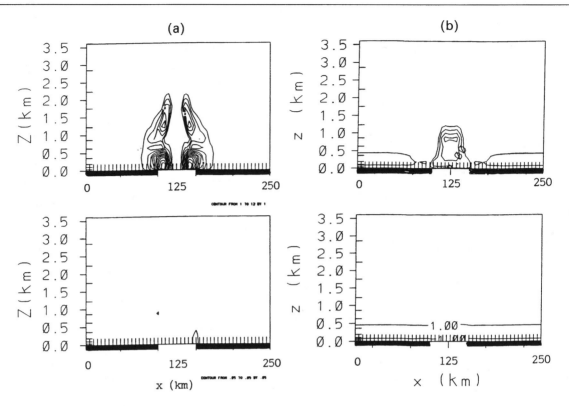

Fig. 11 Same as Fig. 3, but at 3 p.m. only for (i) a deforested area watered to a soil moisture content of 25% of the soil water content at field capacity, and (ii) a deforested area watered to a soil moisture content of 50% of the soil water content at field capacity.

4. SUMMARY AND CONCLUSIONS

Recent investigations suggested that mesoscale fluxes, associated with mesoscale circulations generated by landscape discontinuities, are likely to affect significantly the mean atmospheric variables predicted in large-scale models, e.g., GCMs. Pielke et al. (1991) and Avissar and Chen (1993) emphasized that these mesoscale fluxes may be much more important than turbulent fluxes. While turbulent fluxes are represented in current large-scale models (yet in a questionable way, as was illustrated in this study), mesoscale fluxes have been ignored.

In an attempt to remedy this deficiency, Avissar and Chen (1993) suggested separating the instantaneous atmospheric variables prognosed in large-scale models into large-scale mean, mesoscale perturbation, and turbulent perturbation. Based on this separation, they developed a set of prognostic equations for large-scale mean atmospheric variables, which include the contribution of mesoscale fluxes, as well as the contribution of turbulent fluxes. In addition, they developed prognostic equations for these mesoscale fluxes, which indicate a closure problem. Thus, it is necessary to develop a parametrization for these mesoscale fluxes.

For this purpose, one needs to identify variables that are well related to the mesoscale fluxes, as well as to the physical processes that generate these fluxes. Moreover, we must be able to estimate correctly these variables (or parameters) independently from lower order variables. We believe that the mesoscale kinetic energy (MKE), which is obviously well related to the mesoscale circulations generated by landscape discontinuities, could serve this purpose.

Using a mesoscale atmospheric model to simulate mesoscale circulations generated by local deforestation of the Amazonian region, we diagnosed the spatial and temporal variation of MKE and of the mesoscale heat fluxes within a domain corresponding to a single grid element of a GCM. We analyzed qualitatively the relationship between MKE and the mesoscale fluxes and compared them to turbulence kinetic energy (TKE), and turbulent heat fluxes. Since MKE is relatively well related (though nonlinearly) to the processes that have a major impact on the development of the mesoscale circulations (and, as a result, of the mesoscale fluxes), it is reasonable to assume that this variable could be used for the parametrization of mesoscale fluxes in large-scale models.

Unfortunately, there is currently almost no observational data set that could be used to study the relationship between MKE, landscape discontinuities, and mesoscale fluxes, and, therefore, provide the information necessary to develop a good parametrization. Clearly, to produce such a data set will be complicated and costly. Thus, the development of this

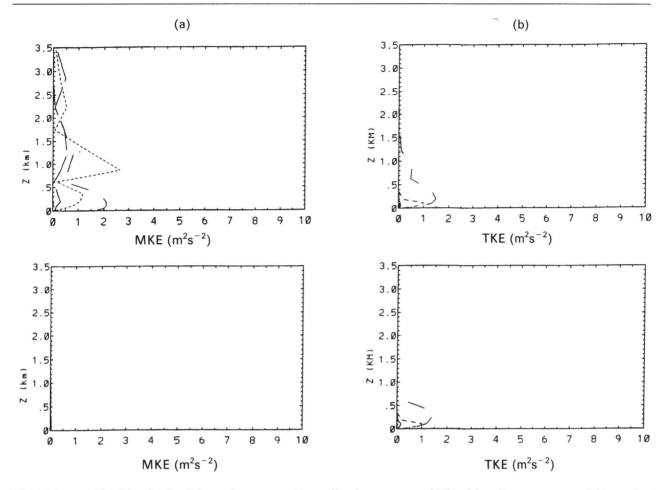

Fig. 12 Same as Fig. 4, but for (i) a deforested area watered to a soil moisture content of 25% of the soil water content at field capacity, and (ii) a deforested area watered to a soil moisture content of 50% of the soil water content at field capacity.

parametrization will need to rely extensively on state-of-the-art numerical modeling analyses. These analyses are likely to help identify the major processes that will eventually need to be validated against field observations and, as a result, will certainly contribute to designing such field experiments.

In order to provide more accurate analyses, the mesoscale atmospheric model to be used for the study should be able to simulate mesoscale perturbations as they are defined in our set of prognostic equations. Thus, the reference atmospheric conditions, which were held constant in the model we used in the present study, will need to be modified to represent large-scale mean conditions that can be updated according to the contribution of the mesoscale and turbulent fluxes.

Clearly, subsequent investigations are necessary to quantify the relationship between mesoscale circulations, mesoscale fluxes, and MKE, under various atmospheric background (large-scale) conditions, and for various landscape discontinuities. Furthermore, since turbulent heat fluxes are affected by mesoscale fluxes, it is necessary to improve the current parametrization of turbulence in large-scale models. We believe that MKE could be used for this purpose as well.

Our simulations indicate that local deforestation in Amazonia tends to increase vertical moisture heat flux in the PBL by the combined effect of turbulence and mesoscale circulations. As a result, local deforestation seems to increase cloudiness and precipitation. Clearly, these results are in contradiction with the prediction of current GCMs, which indicate a reduction of precipitation due to large-scale deforestation. Thus, in order to clarify this issue, we suggest using a mesoscale model nested within a GCM, which could simulate a progressive deforestation of the Amazonian region. Alternatively, similar results should be predictable with a GCM that would be based on the set of prognostic equations and the parametrization suggested in our study.

Finally, it should be mentioned that this paper was accepted for publication in 1992. While the major issues discussed in this paper are still relevant, a few papers on the parametrization of mesoscale fluxes in large-scale atmospheric models have since then been published in the literature. This includes the papers of Chen and Avissar (1994a, b) and Lynn et al. (1995a, b).

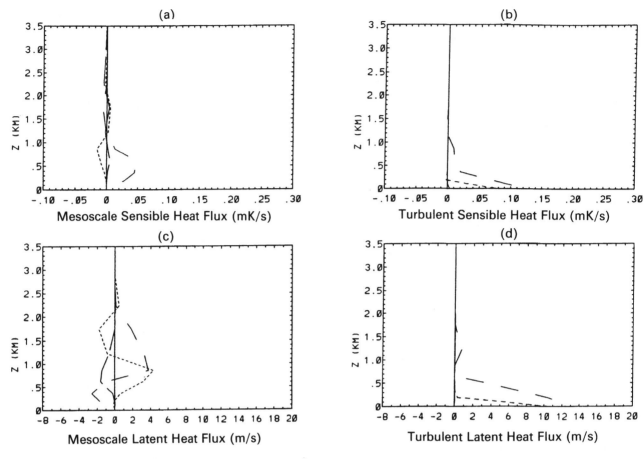

Fig. 13 Same as Fig. 5, but for a deforested area watered to a soil moisture content of 25% of the soil water content at field capacity.

ACKNOWLEDGEMENT

This research was supported by the National Science Foundation (NSF) under Grants ATM-9016562 and EAR-9105059, and by the National Aeronautics and Space Administration under Grant NAWR-28152. Simulations were produced with the Regional Atmospheric Modeling System (RAMS) using the National Center for Atmospheric Research (NCAR) supercomputer facilities. RAMS was developed under the support of NSF and the Army Research Office (ARO). NCAR is supported by NSF. The authors wish to thank Dr. C.A. Nobre for providing the satellite image of local deforestation in the Brazil Amazonian Tropical Forest (Fig. 1 of this paper).

REFERENCES

ANDRÉ, J.-C., G. DE MOOR, P. LACARRÈRE, G. THERRY, and R. DU VACHAT, 1978. Modeling the 24-hour evolution of the mean turbulent structures of the planetary boundary layer. J. Atmos. Sci., 35:1861–1883.

AVISSAR, R., 1991. A statistical-dynamical approach to parametrize subgrid-scale land-surface heterogeneity in climate models. Surv. Geophys., 12:155–178.

AVISSAR, R., 1992. Conceptual aspects of a statistical-dynamical approach to represent landscape subgrid-scale heterogeneities in atmospheric models. J. Geophys. Res., 97:2729–2742.

AVISSAR, R. and F. CHEN, 1993. Using the mesoscale kinetic energy (MKE) for the parametrization of subgrid-scale (mesoscale) processes in large-scale atmospheric models. Part I: definition and analysis. J. Atmos. Sci., 50:3751–3774.

AVISSAR, R. and Y. MAHRER, 1988. Mapping frost-sensitive areas with a three-dimensional local scale numerical model. Part I: Physical and numerical aspects. J. Appl. Meteor., 27:400–413.

AVISSAR, R. and R.A. PIELKE, 1989. A parametrization of heterogeneous land surface for atmospheric numerical models and its impact on regional meteorology. Mon. Wea. Rev., 117:2113–2136.

AVISSAR, R. and M.M. VERSTRAETE, 1990. The representation of continental surface processes in atmospheric models. Rev. Geophys., 28:35–52.

BOUGEAULT, P., B. BRET, P. LACARRÈRE, and J. NOILHAN, 1991. An experiment with an advanced surface parametrization in a mesobeta-scale model. Part II: the 16 June 1986 simulation. Mon. Wea. Rev., 119:2374–2392.

CHEN, F. and R. AVISSAR, 1994a. Impact of land-surface moisture variability on local shallow convective cumulus and precipitation in large-scale models. J. Appl. Meteor., 33:1382–1401.

CHEN, F. and R. AVISSAR, 1994b. Impact of land-surface wetness heterogeneity on mesoscale heat fluxes. J. Appl. Meteor., 33:1323–1340.

COLLINS, D. and R. AVISSAR, 1994. An evaluation with the Fourier Amplitude Sensitivity Test (FAST) of which land-surface parameters are of greatest importance for atmospheric modelling. J. Climate, 7:681–703.

DEARDORFF, J.W., 1974. Three-dimensional numerical study of the height and mean structure of a heated planetary boundary layer. Bound.-Layer Meteorol., 7:81–106.

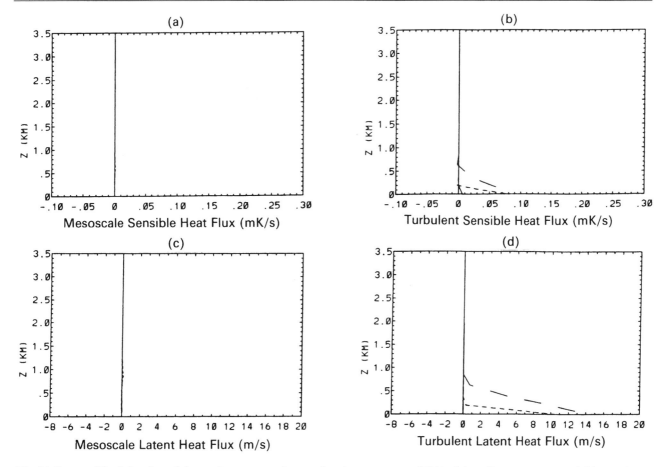

Fig. 14 Same as Fig. 5, but for a deforested area watered to a soil moisture content of 50% of the soil water content at field capacity.

DICKINSON, 1987. (Ed.), The Geophysiology of Amazonia: Vegetation and Climate Interactions, Wiley & Sons, 526 pages.

DICKINSON, R.E. and A. HENDERSON-SELLERS, 1988. Modelling tropical deforestation, a study of GCM land-surface parametrizations, Quart. J. R. Met. Soc., 114:439–462.

ENTEKHABI, D. and P.S. EAGLESON, 1989. Land-surface hydrology parametrization for atmospheric general circulation models including subgrid-scale spatial variability. J. Climate, 2:816–831.

FAMIGLIETTI, J.S. and E.F. WOOD, 1991. Evapotranspiration and runoff from large land areas: land surface hydrology for atmospheric general circulation models. Surv. Geophys., 12:179–204.

HENDERSON-SELLERS, A., R.E. DICKINSON, and M.F. WILSON, 1988. Tropical deforestation, important processes for climate models, Climatic Change, 13:43–67.

LI, B. and R. AVISSAR, 1994. The impact of variability of landscape characteristics on the variability of land-surface heat fluxes. J. Climate, 7:527–537.

LYNN, B., D. RIND, and R. AVISSAR, 1995a. The importance of mesoscale circulations generated by subgrid-scale landscape heterogeneities in general circulation models. J. Climate, 8:191–205.

LYNN, B., F. ABRAMOPOULOS, and R. AVISSAR, 1995b. Using similarity theory to parameterize mesoscale heat fluxes generated by subgrid-scale landscape discontinuities in GCMs. J. Climate, 8: in press.

MAHFOUF, J.F., E. RICHARD, and P. MASCART, 1987. The influence of soil and vegetation on the development of mesoscale circulation. J. Climate Appl. Meteor., 26:1483–1495.

MASCART, P., O. TACONET, J.-P. PINTY, and M. BEN MEHREZ, 1991. Canopy resistance formulation and its effect in mesoscale models: a HAPEX perspective. Agric. For. Meteorol., 54:319–351.

MELLOR, G.L. and T. YAMADA, 1982. Development of a turbulence closure model for geophysical fluid problems. Rev. Geoph. Spa. Phy., 20:851–875.

PIELKE, R.A., G. DALU, J.S. SNOOK, T.J. LEE, and T.G.F. KITTEL, 1991. Nonlinear influence of mesoscale land use on weather and climate. J. Climate, 4:1053–1069.

PIELKE, R.A., W.R. COTTON, R.L. WALKO, C.J. TREMBACK, M.E. NICHOLLS, M.D. MORAN, D.A. WESLEY, T.J. LEE, and J.H. COPELAND, 1992. A comprehensive meteorological modeling system – RAMS. Meteor. Atmos. Phys., 49:69–91.

SEGAL, M., R. AVISSAR, M. MCCUMBER, and R.A. PIELKE, 1988. Evaluation of vegetation effects on the generation and modification of mesoscale circulations. J. Atmos. Sci., 45:2268–2292.

SHUKLA, J., C. NOBRE, and P. SELLERS, 1990. Amazon deforestation and climate change. Science, 247:1322–1325.

STULL, R.B., 1988. An Introduction to Boundary Layer Meteorology. Kluwer Academic Publishers, Dordrecht, The Netherlands, 666 pp.

THERRY, G. and P. LACARRÈRE, 1983. Improving the eddy kinetic energy model for planetary boundary layer description. Bound.-Layer Meteor., 25:63–88.

YAMADA, T. and G.L. MELLOR, 1975. A simulation of the Wangara atmospheric boundary layer data. J. Atmos. Sci., 32:2309–2329.

A hierarchical approach to the connection of global hydrological and atmospheric models

G.W. KITE[1], E.D. SOULIS[2] and N. KOUWEN[2]

[1] *National Hydrology Research Institute,*

11 Innovation Blvd., Saskatoon, Saskatchewan S7N 3H5,

Canada

[2] *Dept. of Civil Engineering, University of Waterloo,*

Waterloo, Ontario N2L 3G1,

Canada

ABSTRACT Hydrology has important contributions to make to the development of global models by providing independent calibration of land surface components of global circulation models (GCMs), by validation of GCM outputs at the basin level through comparison of recorded and simulated streamflows and by examining the implications of climatic change scenarios on water resources. Most importantly, hydrological models are necessary to provide the lateral links needed to close the land surface/boundary layer feedback loops. At grid scales, lateral water fluxes significantly affect soil moisture availability for evapotranspiration. At regional scales, accumulated runoff determines the freshwater inflows to the oceans which drives the sea-ice cover and ocean salinity flows. A hierarchy of hydrological models (HHM) has been developed using a 'grouped response unit' (GRU) approach to link process parameters to land cover, basin topography and the areal extent of climatological phenomena. The GRU allows land-use/land-cover to vary from element to element within a unit. Analyses of data suggest a grid element in the order of 10 km x 10 km is appropriate for hydrological modelling. Temporal resolution is accommodated by using three hydrological models operating at hourly, daily and monthly time scales.

1. INTRODUCTION

Climatologists have been developing and operating atmospheric circulation models for several decades with almost no input from hydrologists. It might be asked why hydrologists should become involved at this stage; why should the integration of hydrological and atmospherical models be considered at all? Hydrological models have had a quite different background and objective than atmospherical models and there has been very little in common. The impetus for the present efforts at cooperation and coordination comes from the post-industrialization modification of the earth's atmosphere and the anticipated effects of these changes on our lifestyle. Perhaps without this impetus, GCMs would have gone on slowly improving, as before; but people live within

the landphase portion of the hydrological cycle and hydrologists must therefore become involved.

Hydrologists have the opportunity to develop the tools needed both to improve global modelling by including better representations of hydrological processes and to evaluate the effects of global change on water resources. This can only be done by reducing the gaps between the spatial and temporal scales of hydrological models and GCMs.

Hydrological models can contribute to modelling of the global cycle by providing independent estimates of evapotranspiration fluxes back to the atmosphere as well as information on regional variations in water availability for soil moisture, water supply, surface runoff, groundwater and fresh-water inputs to the oceans. By using accurately measured streamflows, hydrological models can be used for validation of GCMs. Since measured streamflow integrates all the processes of the hydrological cycle, it provides a check on the overall output of the GCM. However, because of the differences in time scale between the GCMs and the stream-flow response, the value of such validation is limited to longer time periods. These hydrological contributions all necessitate the computation of lateral water fluxes, which is the strongpoint of conventional hydrological models.

To adapt the existing type of hydrological model to a new role as a macro-scale surface-phase component of a global model will mean major changes. The model must accurately represent state and process variables such as groundwater level and infiltration rates as well as the more usual stream-flow. Whereas the objective of the historical hydrological model has been to simulate streamflow and the model has usually been calibrated to this single variable, the objective of the new models must be to simulate a complete range of state variables and fluxes including energy interchanges with the atmosphere.

A macro-scale hydrological model must be applicable to large and diverse regions without recalibration and must be able to operate at a wide range of time scales. The model must include the effects of topography, land-use/land-cover and local meteorological variations (for example, orographic precipitation effects). The spatial coverage of the model must range from the smallest hydrological element to the size of the grids used by GCMs. Experience in developing large-scale models (cf. Becker & Nemec, 1987; Kouwen et al., 1990; Kite, 1991) indicates that, from many viewpoints, a grid size of 10–12 km is appropriate for regional hydrological modelling. Data for land-use/land-cover are available at a resolution of 30 m, or finer, but for practical purposes have been aggregated up to the 10–12 km scale (Ott et al., 1991). Channel hydraulics for hydrological modelling may be performed adequately using a reach length of 10 km (Tao & Kouwen, 1989).

The time resolution of a macro-scale hydrological model depends on both the internal requirements of the model simulations and on the need to transfer data to and from a GCM. Channel routing is probable the most time-critical component of the hydrological model varying from hourly routing using kinematic wave theory, to daily water transfer using non-linear reservoirs and monthly water balance calculations using linear reservoirs. The model must include a mechanism for aggregating and disaggregating data to the typical GCM requirement of data at 15 minute intervals over a period of years.

To accommodate these complex requirements, the macro-scale hydrological model must be physically-based with appropriate representations of the land-phase processes. It would be possible to develop physically-based models that are lumped as, for example, the current work on soil/

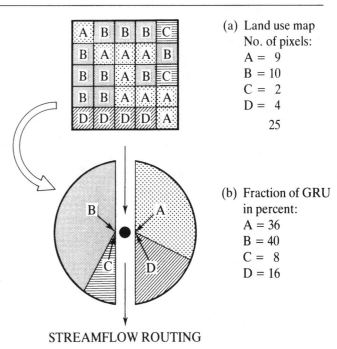

(a) Land use map
No. of pixels:
A = 9
B = 10
C = 2
D = 4
      25

(b) Fraction of GRU in percent:
A = 36
B = 40
C = 8
D = 16

STREAMFLOW ROUTING

Fig. 1 An example of land-use pixels converted to a grouped response unit (GRU).

vegetation/atmosphere/transfer schemes (SVATs), (cf. Avissar & Verstraete, 1990). But, given the distributed nature of precipitation and the non-homogeneity of most basins, a more realistic approach may be to use a physically-based distributed hydrological model. Each of the physically-based components of the distributed model should be verifiable individually. The key to the development of a physically-based macro-scale hydrological model is that the physical descriptions must be appropriate; there is no sense in trying to apply too detailed a model. The data needed will not be available at the macro-scale and the micro-scale physics may not be applicable anyway (Becker & Nemec, 1987).

Most existing distributed hydrological models are based on a grid-square arrangement which allows for the definition of physical parameters in terms of fixed direction and the easy importation of data from remotely-sensed sources. However, using grid squares may mean unnecessary computation if many grid squares have the same properties and if flows are routed from square to square. The finer the grid size used, the more inhomogeneity will be found and the more computation will be needed. Given the uncertainties in meteorological inputs to hydrological models, a point will be reached at which this becomes unreasonable.

The technique used in the hierarchical hydrological model (HHM) is to develop macro-scale hydrological modelling as a hierarchy of three models using the concept of the grouped response unit (GRU). A GRU (Fig. 1) is made up of computational elements consisting of groups of homogeneous pixels or grid squares which have similar characteristics (Kouwen et al., 1990). In the hierarchical model, the GRUs combine groups of grid squares based on land-use characteristics such as grassland, coniferous forest, etc.. The similar grid squares need not be contiguous and the GRU size may be selected to suit routing and meteorological considerations.

The fluxes from the various land-use areas within the GRU are combined at the GRU level and then routed to the next GRU (Fig. 1). Tao & Kouwen (1989) have shown that GRUs can be used even if the component grid squares are at the same elevation, are non-contiguous and are differentiated only by land-use or land-cover. This makes it possible to use the GRU approach in heterogeneous basins and in mountainous basins where unique land-covers at specific elevation ranges are simply a special case. For modelling purposes, the locations of specific land-uses within a GRU are not important – only the relative areas of each land-use are of interest. This information may be readily derived from satellite.

2. THE HIERARCHICAL HYDROLOGICAL MODEL (HHM)

The GRU has been used to develop a hierarchy of three hydrological models which operate at different spatial and temporal scales using common components and a common database to provide a continuum from the GCM scale to the scale of the hydrological/hydraulic application (Fig. 2). The multi-level approach to macro-scale hydrological modelling has been proposed before, e.g. Becker & Nemec (1987), but only within the same model.

2.1 Small-scale component

The approach of representing basin heterogeneity using GRUs was first demonstrated in an event-based flow forecasting model (Kouwen et al., 1990). This component of the HHM is an integrated set of programs to forecast flows for basins with response times ranging from one hour to several weeks. The emphasis is on making optimal use of remotely sensed data; real-time rainfall data from weather RADAR and satellite-derived land-use/land-cover and snowcover data can be directly incorporated into the hydrological model.

The hydrological simulation model is self-calibrating in terms of initial conditions and basin parameters. The model was first derived for urban storm water modelling where the contributions from pervious and impervious areas are calculated separately and added prior to sewer and streamflow routing. This concept of separate contributions was then extended to agricultural areas and broadened into the GRU

concept. As with the other HHM components, using GRUs means that the model can be transferred to many basins without recalibration. The model is being used as a stand-alone forecasting model and for research into rain-on-snow flooding using C-band SAR images. The model has been developed over many years and has been applied extensively in the southern Ontario area.

2.2 Intermediate-scale component

The intermediate-scale component of the hierarchical hydrological model acts both as a basin model in its own right, for space scales of the order 100–10 000 km^2 and a time scale between one day and one month. It acts also as a filter connecting the top and bottom level components of the HHM, feeding data on state variables and fluxes up and down the scale. The model was originally developed as an alternative to existing complex models for Canadian basins (Kite, 1975). It has since been reworked to use remotely-sensed data from satellite (Kite, 1989; Slough & Kite, 1992) and to make use of the GRU concept (Kite & Kouwen, 1992).

The model uses daily precipitation, evaporation and temperature data to simulate streamflow. In addition, recorded streamflow data may be used for calibration. Optionally, satellite-based cloud and snowcover data and recorded snowcourse data may be used (Kite, 1989). A PC database for the model combines the ground-based point data with areally-distributed physiographic, soil, land-use data and satellite data. Daily temperature and precipitation are derived for each GRU by weighted averaging of values from up to five climate stations; missing data are accounted for in the averaging process. Actual evapotranspiration E_a is derived using Morton's (1983) complementary relationship areal evaporation (CRAE) model as:

$$E_a = 2E_w - E_p \qquad (1)$$

where E_w is evapotranspiration under wet conditions and E_p is potential evapotranspiration. CRAE uses mean air and dewpoint temperatures and hours of bright sunshine to calculate E_p using an iterative solution of the energy-balance and vapour transfer equations for dry conditions.

Daily snow and cloud cover may be estimated from a supervised classification of NOAA AVHRR images using a UTM grid square over the basin. Snow water equivalent data may be used as averages of the figures reported from available snow courses or may be derived from passive microwave data from DMSP SSM/I satellite (Slough & Kite, 1992). Land-use data are derived from a multi-spectral classification of Landsat MSS images (Kite, 1989). Table 1 summarizes the use of satellite data within the model. Duchon et al. (1992) used a similar grouping of land-uses

Top Level (input/
output to GCM's)

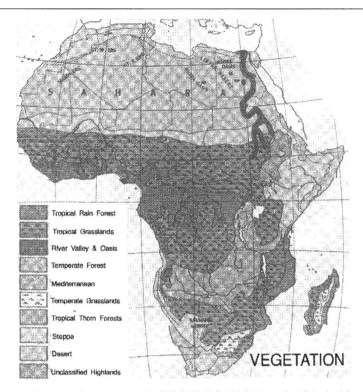

Intermediate Level
(watershed model)

Bottom Level
(hydraulic routing)

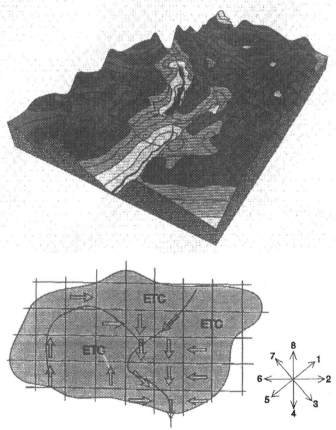

Fig. 2 The NHRI/Waterloo hierarchical hydrological model (HHM) showing the three model levels.

Table 1. *Satellite data used in the basin model.*

Satellite & sensor	Resolution	Frequency	Cost Can $/image	Use in the model
Landsat MSS	80 m	1 in 8 days	800	Land-use
NOAA AVHRR	1 km	2 per day	25	Snow & cloud cover
DMSP SSM/I	25 km	1 per day	20	Snow water equivalent

classified from Landsat images to apply a water balance model to a 538 km² basin in Oklahoma.

The model is applied sequentially to a matrix of GRUs and land-uses for each time unit. Each GRU consists of groups of 12×12 km grid squares based on land-use characteristics (Fig. 3). The GRU is represented in the model by three non-linear reservoirs, one for snowpack, one for a rapid response (may be considered as a combined surface storage and top soil layer storage) and one for a slow response (may be considered as groundwater). The three reservoirs have specified initial contents; there is a maximum depression storage and a maximum allowable depth for slow storage. The model has a total of 14 parameters and operates on a daily time interval. Table 2 lists some typical parameter values for four land-uses in the Kootenay Basin of the Rocky Mountains. The logical relationships between the parameter values and the land-uses is clear from this table.

If the daily mean temperature is above a specified critical temperature then the precipitation is assumed to be rainfall and enters the rapid storage reservoir from where it may percolate to the slow storage reservoir at a rate governed by the Green-Ampt equation (Philip, 1954):

$$\frac{dF}{dt} = K_s \left[1 + \left[\frac{(m - m_o)(C + P(8))}{F} \right] \right] \qquad (2)$$

where F (mm) is the total depth of infiltrated water, t is the time (days), K_s is the saturated conductivity (mm/day), m (mm) is the moisture storage of the soil averaged over the depth to the wetting front, m_0 (mm) is the initial moisture storage, C (mm) is the capillary pressure head at the wetting front and $P(8)$ (mm) is the depth of water on the soil surface.

In practice, the effective porosity is assumed to be 0.33 and the change in moisture storage is calculated as:

$$m - m_o = 0.33[1 - D_s/P(9)] \qquad (3)$$

where D_s (mm) is the depth of the slow reservoir and $P(9)$ is the maximum allowable capacity (mm) of the slow reservoir. Hence the percolation is

$$\frac{dF}{dt} = K_s[1 + [0.33[1 - D_s/P(9)]*[D_r + \phi]/D_s]] \qquad (4)$$

Table 2. *Parameter values, for four types of land-use in the Kootenay Basin of the Rocky Mountains.*

Parameter	Bare Ground	Coniferous Forest	Grass-land	Swamp/ lake
Init. snow water equivalent (mm)	98	94	44	19
Init. surface storage (mm)	0	0	0	0
Init. groundwater storage (mm)	23	33	16	0
Snowmelt > runoff	0.001	0.001	0.001	0.024
Surface roughness	0.4	0.3	0.5	1.0
Groundwater > runoff	0.013	0.011	0.017	0.020
Saturated permeability (mm/day)	18	18	15	11
Depression storage (mm)	5	12	9	0
Max. groundwater (mm)	32	240	87	67
Precip. factor	0.61	0.71	1.25	1.37
Snowmelt temperature (°C)	9.7	7.8	5.4	3.6
Snowmelt (mm/°C day)	0.52	2.2	2.0	0.63
Ratio volume/temperature	0.11	0.49	0.89	1.0
Moving average (days)	4	4	3	2

where D_r (mm) is the depth of the rapid reservoir and ϕ (mm) is the (suction) head defined as

$$\phi = 250[-\log(K_s/86\,400)] + 100 \qquad (5)$$

If the content of the rapid storage reservoir is greater than the depression storage then it may also be depleted to runoff, Q_s, at a rate based on the Manning equation:

$$Q_s = [D_r - P(8)]^{1.67} S_0^{0.5} A \frac{1}{n} \qquad (6)$$

where S_0 is the average overland slope, A is the area of the land-use (km²) and n is a roughness parameter for overland flow.

At the same time, any snowpack is depleted to snowmelt depending on the specified snowmelt rate and the daily air temperature. Snowmelt is proportioned between the rapid storage reservoir and the slow storage reservoir. Snowmelt may, optionally, be modified by the areal extent of snow-cover observed from satellite (Kite, 1989).

If, on the other hand, the daily mean temperature is below the critical value, then the precipitation is assumed to be snowfall and is added to any existing snowpack. There would be no snowmelt in this case, but the rapid storage would be allowed to infiltrate and to runoff as before.

Slow storage contributes to streamflow at a rate depending on the content of the reservoir, and on the temperature and is passed through a moving average filter of variable length. Areal evapotranspiration is satisfied, if possible, first from the snowpack and then from rapid storage and, finally, from slow storage.

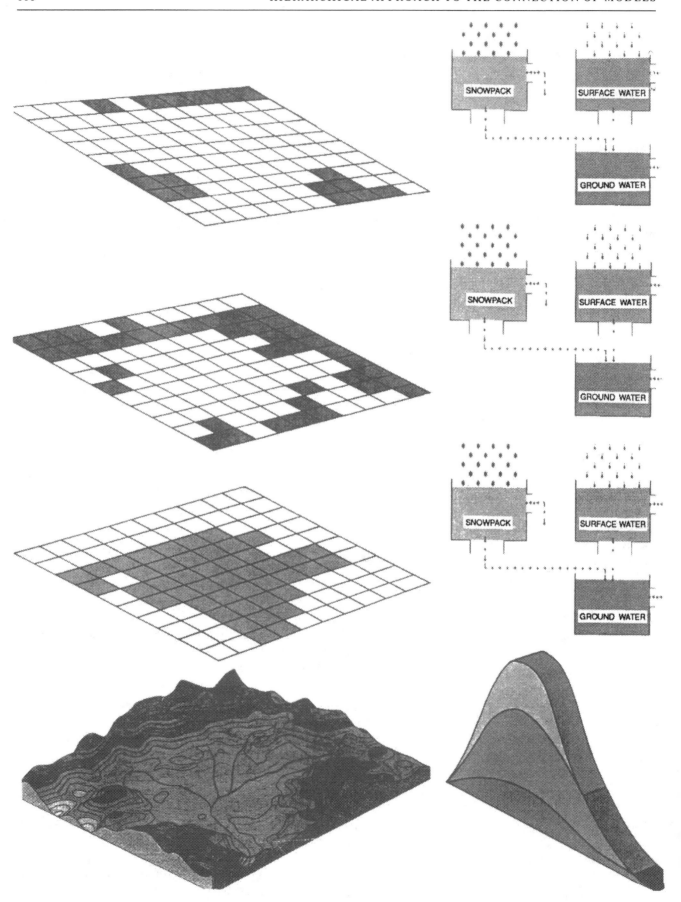

Fig. 3 Using the grouped response unit (GRU) in the intermediate-scale component of the hierarchical hydrological model (HHM).

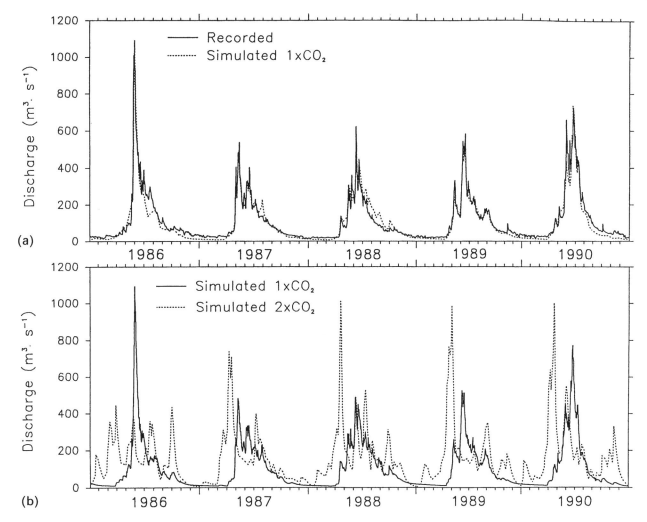

Fig. 4 (a) Comparison of simulated and measured streamflow, Kootenay Basin, 1986–1990. (b) Comparison of streamflow simulation under $1 \times CO_2$ and $2 \times CO_2$ scenarios (after Kite, 1992).

Fig. 4(a) shows a comparison of the recorded and calculated hydrographs for 1986–1990 for the $7100 \, km^2$ catchment of the Kootenay River at Skookumchuck, in British Columbia, to indicate the close simulation that is possible with this model.

2.3 Large-scale component

The third component of the hierarchical hydrological model is the top-level model operating at a coarse grid scale. This is an adaptation of the GHEIS model (Shawinigan Consultants, 1983) and ESMMap model (Solomon et al., 1991) used in Canada, England, South America and West Africa. GHEIS is a parametric model designed to simulate monthly or annual flows for large regions up to continental scale by relating precipitation, temperature and streamflow to physiographic and land-use characteristics.

GHEIS operates as a water balance model, using Turc's formula (Gray, 1970) to estimate evapotranspiration from temperature and precipitation, and extrapolate the climatic variables using physiographic data. The physiographic characteristics used include elevation, slope, azimuth, barrier height, shield effect and distance to the sea. The same land-use characteristics are used as in the other two HHM components. This model can be adapted to use the precipitation and temperature outputs from GCM grid points (Fig. 5).

The three levels of model, flow routing, basin and water balance are then combined into the HHM using a common databank. The advantage of developing a hierarchy of models all using GRUs based on land-use characteristics is that they can be used to simulate the effects of changing land-use. For example (see Fig. 4(b)), by using the anticipated changes in land-use within a basin, the intermediate scale basin model has been used to simulate the effects of a doubling of CO_2 on streamflow (Kite, 1992).

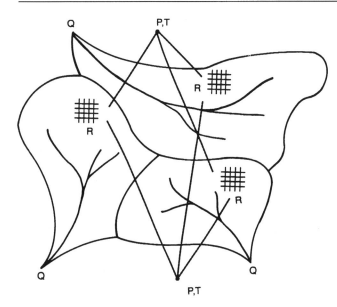

JOINT INTERPOLATION OF P,T, AND R

Fig. 5 Large-scale component of the hierarchical hydrological model (HHM) showing the joint interpolation of precipitation *P*, temperature *T* and runoff *R*.

CONCLUSIONS

A macro-scale hydrological model must represent the physics of catchment hydrology but still remain computationally efficient. The grouped response unit (GRU) combines groups of pixels or grid squares with similar land-use characteristics. The pixels or grid squares do not need to be contiguous and the element size may be selected to suit model scale considerations. Using GRUs means that runoff from different hydrological components of the basin may be computed separately and then added together.

The GRU concept has allowed the development of a hierarchy of hydrological models at different scales, using common components and a common database. Each model uses similar process descriptions and uses GRUs made up of different land-use elements. The largest scale component is based on a well-established square-grid model, the intermediate scale component uses a semi-distributed basin model capable of using data from Landsat, NOAA and DMSP satellites and the smallest scale component uses a flow forecasting model able to use satellite and RADAR data.

The models are now used individually; the next step is to link the models through the common database. Testing will be carried out in the Mackenzie Basin within the Canadian contribution to the GEWEX Continental-Scale International Project (WMO, 1991).

Because the GRUs are based on land-use characteristics, the HHM and its component models may be used to investigate changing land-use patterns such as might be caused by climatic change.

REFERENCES

AVISSAR, R., & VERSTRAETE, M.M. (1990). The representation of continental surface processes in atmospheric models. Rev. Geophys., 28, 35–52.

BECKER, A. & NEMEC, J. (1987). Macroscale hydrologic models in support to climate research. In: The influence of climate change and climatic variability on the hydrologic regime and water resources, IAHS Publ. no. 168, 431–445.

DUCHON, C.E, SALISBURY, J.M., LEE WILLIAMS, T.H. & NICKS, A.D. (1992). An example of using Landsat and GOES data in a water budget model. Wat. Resour. Res., 28(2), 527–538.

GRAY, D.M. (1970). Handbook on the principles of hydrology. NRC, Ottawa.

KITE, G.W. (1975). Performance of two deterministic models. In: Application of mathematical Models in Hydrology and Water Resources Systems (Proc. of the Bratislava Symposium, September 1975). IAHS Publ. no. 115, 136–142.

KITE, G.W. (1989). Hydrological modelling with remotely sensed data. Proc. 57th Annual Western Snow Conference. Ft. Collins, Co., April 18–20, 1–8.

KITE, G.W. (1991), A watershed model using satellite data applied to a mountain basin in Canada. J. Hydrol., 128, 157–169.

KITE, G.W. (1992). Applying a climatic change scenario to a semi-distributed watershed model. Proc. 60th Annual Western Snow Conference. Jackson Hole, Wyoming, April 13–16.

KITE, G.W. & KOUWEN, N. (1992): Watershed modelling using land classifications, Wat. Resour. Res., 28, 12, 3193–3200.

KOUWEN, N., SOULIS, E.D. & PIETRONIRO, A. (1990). Enhancing rainfall-runoff modelling of mixed land-use/land-cover areas with remote sensing. In: Applications of Remote Sensing in Hydrology ed. G.W. Kite & A. Wankiewicz (Proc. NHRI Symposium No. 5) NHRC, Saskatoon, 94–108.

MORTON, F.I. (1983). Operational estimates of areal evapotranspiration and their significance to the science and practice of hydrology. J. Hydrol., 66, 1–76.

OTT, M., Z. SU, SCHUMANN, A.H. & SCHULTZ, G.A. (1991). Development of a distributed hydrological model for flood forecasting and impact assessment of land-use change in the international Mosel River Basin. In: Hydrology for the Water management of Large River Basins (Proc. of the Vienna Symposium, August, 1991), IAHS Publ. no. 201, 183–194.

PHILLIP, J.R. (1954). An infiltration equation with physical significance. Soil Science, 77(1), 153–157.

SHAWINIGAN CONSULTANTS INC & S.I. SOLOMON AND ASSOCIATES LTD. (1983). A square grid study of the Amazon River Basin. World Meteorological Organization Report No. 8, Project BRA/72/010, WMO/UNDP, Geneva.

SLOUGH, K. & KITE, G.W. (1992) Remote sensing estimates of snow water equivalent for hydrological modelling applications. Can. J. Wat. Res., 17, No 4, 323–330.

SOLOMON, S.I., CAPONE, C., MOREAU, A., DENGO, M. & LEE, A. (1991). The Geographic, Environmental, Economic and Social Information System (GEnESIS) for artificial intelligence applications. Unpublished paper presented at the XVI General Assembly, European Geophysical Society, Wiesbaden.

TAO, T. & KOUWEN, N. (1989). Remote sensing and fully-distributed modelling for flood forecasting. J. Water Resour. Plng. and Mgmt., ASCE, 115 (6), 809–823.

WORLD METEOROLOGICAL ORGANIZATION (1991) Scientific plan for the GEWEX Continental-Scale International Project. WRCP WMO/TD, WMO, Geneva, 71pp.

Stochastic downscaling of GCM-output results using atmospheric circulation patterns

A. BÁRDOSSY

Institute for Hydrology and Water Resources, University of Karlsruhe,

Kaiserstrasse 12, D-7500 Karlsruhe,

Germany

ABSTRACT A methodology is presented for downscaling GCM-output results to regional scale precipitation using atmospheric circulation patterns. A sequence of observed daily air pressure distributions is used to define circulation patterns. The classification of the circulation patterns is done with the help of a neural network. A multivariate stochastic model describes their link to observed daily precipitation amounts at a number of selected locations. To assess precipitation under changed climate circulation patterns derived from GCM output pressure values are used to condition the stochastic precipitation model. The methodology is demonstrated by results obtained for a selected location (Essen, Germany).

1. INTRODUCTION

Climate change will have a major influence on the hydrological cycle. It is of vital importance to assess the possible impacts as soon as possible, in order to find strategies to adapt to these changes.

The only physically based tools in predicting climate change effects are General Circulation Models (GCM). GCMs deliver meteorological variables in a fine time resolution (30 minutes to a few hours) but in a very coarse spatial grid (200–500 km × 200–500 km). Many climatic parameters like temperature, precipitation, wind, clouds, radiation, snow cover and soil moisture can be simulated by these models.

Unfortunately precipitation, which is the main input in hydrological models, cannot be well modelled by the GCMs. GCM control runs for the present climate indicate that single grid values cannot be taken as representative rainfall amounts of the corresponding area. Fig. 1 shows observed and GCM simulated seasonal precipitation amounts for the block containing Germany. It can be seen that the difference between the observed and $1 \times CO_2$ (present climate control) run is much higher than the climate change signal (difference between $2 \times CO_2$ and $1 \times CO_2$). Furthermore the simulated annual cycle is also incorrect, bringing the highest precipitation amounts in winter instead of the observed maximum

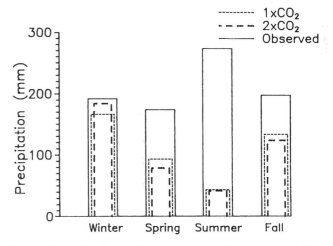

Fig. 1 Observed and simulated seasonal precipitation amounts for the GCM block covering Germany.

in summer. Even if the grid values were accurate, the size of the GCM grid-cells would make a downscaling to hydrologically relevant areas necessary. Nested models (Giorgi and Mearns, 1991) can deliver data in a finer spatial resolution, but as their boundary conditions are given by the coarser GCM it is not certain that they will be accurate enough for hydrological modelling.

GCMs are more accurate on free atmosphere variables such as air pressure. Precipitation can in the temperate zone

119

usually be related to air mass transport. The atmospheric circulation is a consequence of pressure differences and anomalies. Therefore a link exists between air pressure distribution over a large area and the local precipitation. Atmospheric circulation patterns are defined as typical air pressure distributions.

The purpose of this paper is to develop a method for downscaling GCM output results to the regional scale using circulation patterns. Instead of using the inaccurate value of a single GCM block the circulation patterns derived from the pressure values over a set of blocks are used for the downscaling. Observed data can be used to develop a stochastic model conditioned on the circulation patterns. This model can then be used with daily circulation patterns derived from GCM outputs for a changed climate to simulate possible daily precipitation amounts for selected locations. The concept of the methodology has been described by Bárdossy and Caspary (1991).

A similar approach was used by Hay et al. (1991). They defined weather types using air pressure, wind direction and cloud cover, using GCM present climate and doubled CO_2 output. In their model they did not consider any spatial correlation between different stations, and no autocorrelation of the precipitation amount time series. This restricts the application of the model. Wilson et al. (1991) developed a model for daily rainfall at a number of stations depending on the atmospheric circulation patterns. They used air pressure data for recognition of the circulation pattern.

2. ATMOSPHERIC CIRCULATION PATTERNS

Baur et al. (1944) defined a circulation pattern as a mean air pressure distribution over an area at least as large as Europe. Any given circulation type persists for several days (generally at least 3 days) and during this time the main features of the weather remain mostly constant across Europe. After that there is a rapid transition to another circulation type.

Atmospheric circulation patterns can be obtained by the classification of the surface air pressure or the 500 (700) hPa surfaces elevations. The classification is done on a daily basis. There are several classification methods. These include:

- subjective classification (Baur scheme, Lamb's classification);
- k-means clustering (Wilson et al., 1992; Bogárdi et al., 1992);
- fuzzy clustering (Bezdek, 1981);
- fuzzy rule based clustering;
- neural networks.

Subjective classification was the only possible method in the pre-computer era, and it is presently also used. The advan-

tage of this method is that the knowledge and experience of the meteorologist is fully used in the classification, and thus only the main features of the circulation pattern are used. A major disadvantage is that the results cannot be reproduced: two meteorologists might have a different opinion on the classification of a particular day.

For climate change purposes subjective classification of long time series of GCM-outputs requires a day by day evaluation by a meteorologist, which is impracticable. Traditional clustering methods are objective, but they do not profit from the meteorologist's knowledge. For the present study a classification method based on a neural network was selected. This method is a combination of objectivity and subjective meteorological information.

Neural networks (NN) are mathematical models of the brain activity (McCord Nelson and Illingworth, 1991). The basic unit of an NN is the neuron. The functioning of a neuron is described by the transformation of the input signals to an output signal. The state of the neuron depends on the input signals coming from connected neurons. An NN consists of a set of structured neurons. The most common NNs consist of three different types of layer:

- The input layer: these are the neurons which are activated by the input signal coming from outside.
- The hidden layers: these are the neurons which are supposed to perform the transformation of the input to an output.
- The output layer: these are the neurons which provide signals for the outside.

Each neuron of a layer is connected to each neuron of the adjacent layer. The interconnection between the neurons expressed in the form of weights has to be determined with the help of a learning procedure. For this purpose a training set is used. Weights which minimize the difference between the known output of the training set and the calculated output of the NN are determined in this step.

For the circulation pattern classification a neural net consists of 3 layers: a 51 neuron input layer, a 29 neuron output layer and one hidden layer was used. The input to the NN is the normalized 700 hPa surface elevation at 51 selected locations. Fig. 2 shows the location of these points. The output signal is used to decide on the circulation pattern type: the type with the highest signal value is assigned to the input pressure distribution.

The elements of the training set were defined as the mean 700 hPa surfaces of the Baur classification CPs (using 10 years' observations NCAR diamond grid data set) with random perturbations.

The classification recognizes 3 groups of circulations divided into 10 major types and 29 subtypes. The goal of the NN classification was not to reproduce the Baur classifica-

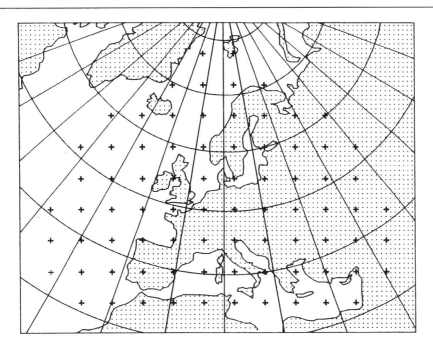

Fig. 2 Location of the 51 points used for the NN classification.

tion but instead to develop a similar objective classification. Classification was performed for three different data sets:

(1) Observed daily 700 hPa data of 1977–1986;
(2) ECHAM-1 GCM 10 years daily $1 \times CO_2$ run, 700 hPa output data;
(3) ECHAM-1 GCM 10 years daily $2 \times CO_2$ run, 700 hPa output data.

For each data set the same NN as obtained from the above defined training set was used.

Table 1 shows the frequency of the different circulation forms obtained by the neural net classification. These preliminary results indicate that there are differences between the observed frequency of circulation patterns and the results of the control run ($1 \times CO_2$). The output of other GCMs can also be classified by means of the same methodology and the performance of the GCMs can also be compared using these frequencies.

3. PRECIPITATION MODELLING

It is obvious that there exists a close relationship between circulation patterns and climatic variables. Bürger (1958) studied the relationship between the atmospheric circulation patterns and mean, maximum and minimum daily temperatures, precipitation amounts and cloudiness using the time series from 1890 to 1950 measured in four German cities (Berlin, Bremen, Karlsruhe and Munich). He found a good correspondence between climatic variables and atmospheric circulation. Lamb (1977) stated that even the highly varying

precipitation is strongly linked to the atmospheric circulation.

Once a circulation pattern classification has been performed, the link to precipitation has to be established. Table 2 shows the precipitation statistics for a few different circulation patterns during the winter at the meteorological station of Essen. It can be seen that there is a considerable difference in the precipitation probabilities ranging from 13.18% (CP HM = High over Central-Europe) to 86.62% (CP NWz = North-West cyclonic). Similar differences can be observed in the other statistics too.

To describe the precipitation at a number of stations a stochastic space-time model for daily rainfall using atmospheric circulation patterns as developed by Bárdossy and Plate (1992) was used. This model uses a transformed conditional multivariate autoregressive **AR**(1) model, with parameters depending on the atmospheric circulation patterns. The model can be briefly described as follows.

Let the daily precipitation amount at time t and point u in the region U be modelled as the random function $Z(t,u)$, $u \in U$. $Z(t,u)$ is related to a simple normally distributed random function $W(t,u)$ through the power transformation relationship:

$$Z(t,u) = \begin{cases} 0 & \text{if } W(t,u) \leq 0 \\ W^\beta(t,u) & \text{if } W(t,u) > 0 \end{cases} \qquad (1)$$

Here β is an appropriate positive exponent. This way the mixed (discrete – continuous) distribution of $Z(t,u)$ is related to a normal distribution. The process $Z(t,u)$ and thus $W(t,u)$ depends on the atmospheric circulation pattern. The expo-

Table 1. *Frequency (%) of the different circulation forms obtained by the neural net classification.*

Circulation form	Data	Winter	Spring	Summer	Fall	Annual
Zonal	Observed	32.04	13.19	10.97	22.75	19.66
	$1 \times CO_2$	19.56	10.11	6.11	16.33	13.03
	$2 \times CO_2$	22.11	12.78	10.56	14.67	15.03
Semi-Meridional	Observed	32.04	22.86	13.98	44.84	28.34
	$1 \times CO_2$	31.89	30.22	38.22	41.56	35.47
	$2 \times CO_2$	33.67	32.11	32.44	38.56	34.19
Meridional	Observed	35.92	63.96	75.05	32.42	52.00
	$1 \times CO_2$	48.56	59.67	55.67	42.11	51.50
	$2 \times CO_2$	44.22	55.11	57.00	46.78	50.78

Table 2. *Precipitation statistics for different circulation patterns (CP) during the winter at Essen (Germany).*

HH = frequency of CP, p_i = probability of precipitation, CT = contribution of CP to total rainfall, WI = wetness index CT/HH, m_i mean daily precipitation on wet days, s_i standard deviation of precipitation amount on wet days.

CP	HH [%]	$p_i(u)$ [%]	CT [%]	WI [−]	m_i [mm]	s_i [mm]
Wa	6.11	49.68	3.92	0.64	3.20	5.01
Wz	17.82	85.62	39.19	2.20	6.38	6.19
SWz	3.47	82.95	6.61	1.91	5.70	5.70
NWz	6.19	86.62	10.92	1.77	5.06	5.47
HM	5.08	13.18	0.94	0.18	3.46	4.03
HB	2.84	29.17	0.36	0.13	1.08	1.38
TrW	3.82	62.89	3.97	1.04	4.10	4.68

Table 3. *Observed and simulated annual and seasonal precipitation amounts (mm) for the meteorological station Essen (Germany).*

Case	Winter	Spring	Summer	Fall	Annual
Observed	227.2	227.4	251.7	203.7	910.0
$1 \times CO_2$	203.0	210.8	197.1	220.8	831.7
$2 \times CO_2$	220.4	221.0	206.5	221.2	863.1

nent β is needed as the distribution of precipitation amounts is usually much more skewed than the truncated normal distribution.

The relationship between $W(t, u)$ and the circulation pattern $\tilde{A}_t$ is obtained through the rainfall process $Z(t, u)$ using eq. (1). The probability distribution of daily rainfall amounts at time t and location u depends on the atmospheric circulation pattern $\tilde{A}_t$. The random process describing $W(t, u)$ is supposed to be a multivariate autoregressive (**AR**(1)) process in the case of a persisting atmospheric circulation pattern. There is no time 'continuity' in $W(t, u)$ if the atmospheric circulation pattern changes at time t.

Model parameters were estimated using the methods described by Bárdossy and Plate (1992). Observed daily precipitation and the NN classification for the time period 1977–1986 were used. Then the model was used for the simulation of daily rainfall for the sequence of daily circulation patterns obtained from the classification of $1 \times CO_2$ and $2 \times CO_2$ GCM output. Table 3 shows the mean annual precipitation amounts for Essen (Germany). It can be seen

that the difference between the observed mean annual precipitation and the $1 \times CO_2$ simulated mean annual precipitation is less than 10%. On the other hand $2 \times CO_2$ simulation yields an increase of precipitation which is less than 5%. Fig. 3 shows the mean seasonal precipitation amounts for the three different cases. Compared to the GCM block results shown in Fig. 1 it can be seen that the agreement between observed and $1 \times CO_2$ simulation is much better in this case. The climate change signal remained in the same order of magnitude as for the entire GCM-block.

4. CONCLUSIONS

GCMs cannot simulate accurate precipitation amounts in the present climate control runs. However, they are more accurate for air pressure. Circulation patterns can be derived from the daily air pressure distributions using neural networks. There is a stochastic link between precipitation and circulation patterns. Using this link local precipitation amounts can be simulated corresponding to the GCM output circulation patterns. These amounts are much closer to the observed values than the GCM output precipitation. Further work is required to:

(1) refine the classification;
(2) use 500 hPa surfaces;

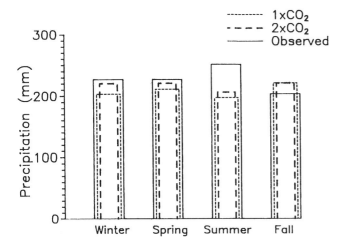

Fig. 3 Observed and downscaled simulated seasonal precipitation amounts for Essen.

(3) check persistence properties of the link under increased temperature;

(4) assess areal precipitation (watershed size).

The presented methodology can be used in connection with the output of any selected GCM. The increasing accuracy of GCMs simulating circulation patterns will certainly improve the accuracy of the local simulation results.

REFERENCES

BÁRDOSSY A. and CASPARY, H. J., 1991: Conceptual model for the calculation of the regional hydrologic effects of climate change, in: Hydrology for the Water Management of Large River Basins, van de Ven, F.H.M., Gutknecht, D., Loucks, D.P. and Salewicz, K.A. (eds.), IAHS Publication 201:73–83 IAHS Press, Wallingford.

BÁRDOSSY, A. and PLATE, E.J., 1992: Space-time model for daily rainfall using atmospheric circulation patterns, Water Resources Research, 28:1247–1259.

BAUR, F., HESS, P. and NAGEL, H., 1944: Kalender der Großwetterlagen Europas 1881–1939. Bad Homburg.

BEZDEK, J.C., 1981: Pattern Recognition with Fuzzy Objective Function Algorithms, Plenum Press, New York.

BOGÁRDI, I., MATYASOVSZKY, I., BÁRDOSSY, A. and DUCKSTEIN, L., 1992: Application of a space-time stochastic model for precipitation using atmospheric circulation patterns, Working Paper, University of Nebraska-Lincoln.

BÜRGER, K., 1958: Zur Klimatologie der Großwetterlagen. Berichte des Deutschen Wetterdienstes Nr. 45, Bd. 6, Offenbach a. Main, Selbstverlag des Deutschen Wetterdienstes.

GIORGI, F. and MEARNS, L.O., 1991: Approaches to the simulation of regional climate change: A review, Reviews of Geophysics, 29: 191–216.

HAY, L.E., MCCABE, G.J., WOLOCK D.M. and AYERS, M.A., 1991: Simulation of precipitation by weather type analysis, Water Resources Research, 27: 493–501.

LAMB, H.H., 1977: Climate, present, past and future. Vol. 2: Climatic history and the future. London, Methuen & Co Ltd, 835 pp.

MCCORD NELSON, M., and ILLINGWORTH, W.T., 1991: A practical guide to neural nets, Addison-Wesley, Reading.

WILSON, L.L., LETTENMAIER, D.P. and WOOD, E.F., 1991: Simulation of precipitation in the Pacific Northwest using a weather classification scheme, Surveys in Geophysics, 12: 127–142.

WILSON, L.L., LETTENMAIER, D.P. and SKYLLINGSTAD, E., 1992: A hierarchical stochastic model of large scale atmospheric circulation patterns and multiple station daily rainfall, Journal of Geophysical Research, 97, ND3: 2791–2809.

Dependencies of spatial variability in fluvial ecosystems on the temporal hydrological variability

H.P. NACHTNEBEL

Institute for Water Resources Management, Hydrology and Hydraulic Engineering,

Universität für Bodenkultur,

A-1190 Vienna, Austria

ABSTRACT Flood plains are regions which are strongly dependent on the hydrological conditions induced by the dynamics of the fluvial water system. These regions also exhibit a broad spectrum of different habitats closely interfaced and of large biological diversity.

The objective of this paper is to describe some relationships between the dynamic characteristics of hydrological variables and the spatial variability in fluvial ecosystems. The interrelationships are not well understood but some hydrological key variables can be identified which drive the biological system. The integration of temporal hydrological characteristics with the spatial variability of some abiotic parameters, such as soil type, yields an indication for the spatial variability in the environment. The links between the hydrological system characterized by the soil moisture budget and the vegetation layers in the riverine forest ecosystem are analysed at the patch scale and the regional scale.

The methodology is applied to a flood plain region located along the Austrian section of the Danube. Due to the implementation of a hydropower scheme the backwater region was separated from the main river by impounding dams located along the old river banks. The temporal pattern of the hydrological variables, including the surface water table and the groundwater table, was completely altered. The changes in both the abiotic and biotic systems were monitored over a rather long time and some links between these systems were studied.

1. INTRODUCTION

Riverine forests constitute habitats exhibiting a high biological diversity. For instance, in the remaining Austrian flood plain forests which extend mainly along the Austrian Danube (Fig. 1), about 12 000 species of flora and fauna are abundant (Gepp, 1985). Due to river training works, river channelization, implementation of hydropower schemes and modified land use the hydrological characteristics in the flood plain forests have been remarkably modified in the last decades (Decamps et al., 1988; Petts et al., 1989). Because of its ecological importance, underlined by the fact that many endangered species have their last habitats in these areas the preservation of riverine forests and wetlands is attracting increasing interest.

Extended flood plain regions (Chang, 1988) are mainly formed by braidening or meandering river systems. The systems of oxbow lakes and ditches resemble old river branches which were subsequently filled by sediments. Within a time horizon of hundreds to thousands of years the temporal variability in the discharge and the succession in the vegetation pattern defined the topography and the distribution of the soil layers. The vegetation did not only react according to the temporal hydrological pattern but also affected erosion processes during flood events by trapping sediments.

The relevant temporal and spatial scales of stream systems and respective habitat structures are given in Fig. 2.

To study changes in the ecological subsystems which become observable within a period of two decades it is reasonable to investigate spatial scales of ten metres to a few hundred metres. The whole flood plain system reacts rather slowly to the modified hydrological boundary conditions. Major changes will become observable at the earliest after decades.

In a case study (Hary & Nachtnebel, 1989) the impacts of

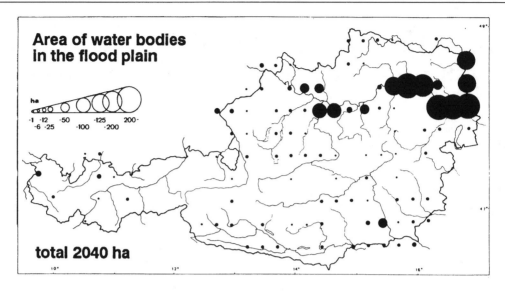

Fig. 1 Regional distribution of flood plain water bodies in Austria (after Gepp, 1985).

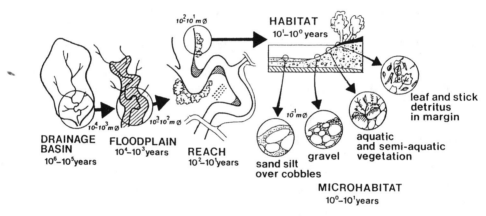

Fig. 2 Hierarchy of stream systems and of habitats (after Frisell et al., 1986).

river impounding on the hydrological and the ecological system were analysed (Fig. 3). While the old flood levees allowed frequent inundation of a three to four kilometres broad riverine forest belt, the new impounding dams which were built together with the hydropower scheme followed the old river banks directly. Thus, these dams separate completely the backwater region from the main river. With respect to the hydrological characteristics the inundation frequency in the former flood plain was drastically reduced and the groundwater system was decoupled from the main river. As an example the time series of the groundwater table of an observation well which is located about 150 m away from the main river is displayed in Fig. 4. Due to the proper technical design of a drainage system along the impervious impounding dams the mean height of the groundwater table could be maintained but the dynamics of the subsurface system was remarkably reduced. It can be concluded from Fig. 4 that the hydropower scheme was put into operation in April 1976.

This paper is organized in four sections. First, the hydrological changes at some locations will be described. Then, the

impacts on the vegetation are analysed and reference is made to the tree layer, the shrub layer and the herb layer. Based on data from several point observations the hydrological changes are regionalized over the whole former flood plain area. The spatial variability of various abiotic parameters such as terrain elevation, soil thickness and groundwater table is integrated and reflected by the soil moisture pattern. Finally, the corresponding impacts of a modified soil water balance on the vegetation are discussed.

2. HYDROLOGICAL REGIME AT SELECTED SITES

In general, the hydrology of the flood plain is controlled by inundation, precipitation and the groundwater table. These variables constitute an input to the soil-vegetation-atmosphere system. The output of the system can be described by the productivity of the biomass or by biological diversity.

To assess the soil water budget at several forest stands a

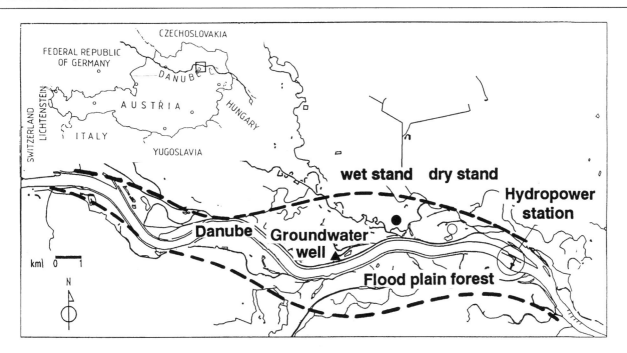

Fig. 3 Hydropower scheme Altenwörth and the flood plain region.

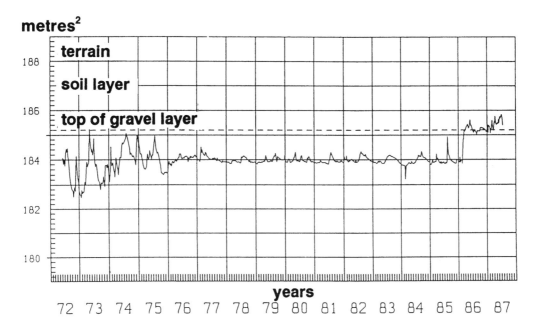

Fig. 4 Time series of the groundwater observations in a well located 150 m from the main river.

numerical model for one-dimensional, vertical unsaturated flow is applied (Feddes et al., 1988). The soil column above the gravel is discretized into several homogeneous layers (Fig. 5). For each layer a set of parameters including thickness d_j, saturated hydraulic conductivity k_{sj} and respective volumetric soil water content θ_j is required. According to Campbell (1985) the unsaturated hydraulic conductivity k_{uj} and the soil water suction Ψ_{nj} are expressed as functions of the volumetric soil water content θ_j and the respective values at saturation, k_s and Ψ_s.

$$k_u = k_s (\theta/\theta_s)^m \qquad (1)$$

$$\Psi_u = \Psi_s (\theta/\theta_s)^{-b} \qquad (2)$$

$$m = 2b + 3 \qquad (3)$$

The vertical flow of water is obtained from the Richards equation describing the change in the volumetric soil moisture budget.

$$\frac{\delta \theta}{\delta t} = \frac{\delta}{\delta z} \left\{ k_u(\theta) \left(\frac{\delta \Psi}{\delta z} + 1 \right) \right\} + Q(z, t) \qquad (4)$$

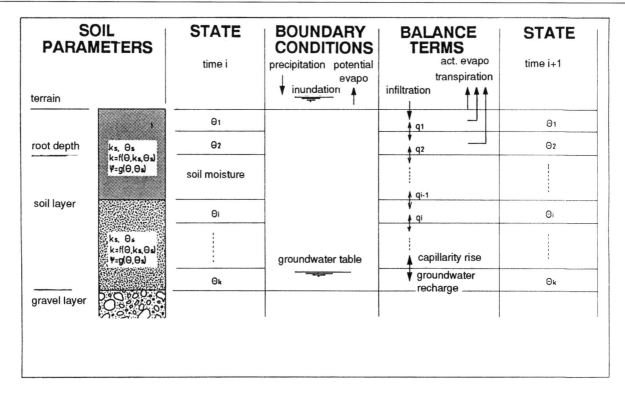

Fig. 5 Simulation model for the water balance of a soil column.

where the sink term $Q(z, t)$ represents the water extraction by plant roots

The boundary conditions are expressed by the depth D of the groundwater table below the surface

$$\theta(z = D) = \theta_s \qquad (5)$$

and a flux term across the soil-atmosphere boundary consisting of

$$Q(z = 0) = Q_o(t) \qquad (6)$$

and

$$Q(z, t) = T_p\{a(\theta_j)|z = d_j\}r(d_j) \qquad (7)$$

$Q_o(t)$ includes the infiltration rate during rainfall or inundation and also the evaporative losses from the top soil layers (Nachtnebel & Haider, 1991). Infiltration during rainfall is dependent on the interception losses which are related to the leaf area index (LAI), i.e. the area of leaves per unit ground area. The potential transpiration T_p depletes the soil moisture storage and acts via the root system on deeper soil layers. A soil extraction factor $a(\theta_j)$ (Markar and Mein, 1987) and a root activity factor $r(z)$ (Feddes et al., 1978; Prasad, 1988) control the sink term down to the bottom of the root zone z_r.

The soil water balance model yields time series of the soil moisture in various layers and also the available soil water in the root zone can be obtained. Due to the fact that only local inundation occurred after implementation of the impounding dams the soil water regime became mainly dependent on

the groundwater level. Two sites, one strongly affected by groundwater and the other one more or less independent of groundwater were selected for detailed hydrological and plant physiological investigations. Based on the soil parameters for each layer the simulated soil water balances in the root zone were compared (Fig. 6) for two sites.

It can be concluded from the observations and the simulation results that the groundwater table responded to the modified boundary conditions within a few weeks (Fig. 4). The soil moisture balance which is additionally dependent on the rainfall shows somewhat delayed reaction but changes become visible within one vegetation period. Even when the groundwater supply to the soil column occurs only for several short periods the decrease in the soil moisture due to water uptake by plants is compensated.

3. EFFECT OF GROUNDWATER TABLE ON PLANTS AT SELECTED SITES

The effects of a change in the soil water budget on vegetation (Reily & Johnson, 1982; Kimmins, 1987) were investigated in the case study by Maier (1989). A plant physiological comparison of the two sites with similar abiotic factors but different groundwater influences showed higher overall productivity of the vegetation on the site with frequent groundwater contact (Table 1). As a typical plant species in the riverine vegetation *Populus Alba aff. canescens* was investi-

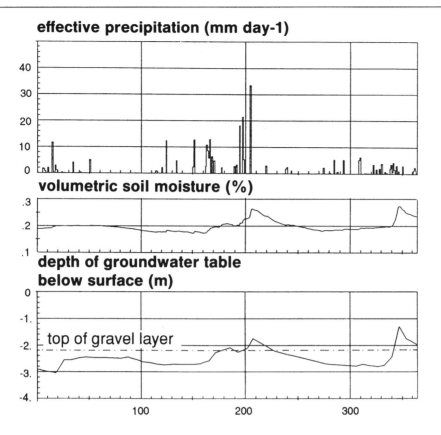

Fig. 6 Effective rainfall, soil moisture and depth of groundwater table for a 'wet site' in the year 1974.

gated in detail. Net photosynthesis per unit leaf area on both sites shows only slight differences. The important loss of assimilates at the site without groundwater contact as compared to the other site is due to the lower LAI. Productivity of the shrub layer (*Cornus Sanguinea*) is also diminished by the lack of groundwater contact.

A lower LAI allows more radiation to pass through the upper part of the treetop. On the site without groundwater contact radiation in the lower part of the treetop is increased by 20% and in the herb layer by 50%.

Also, the reduced LAI resulted in decreased interception losses and hence in an increased throughfall. Therefore, the herb layer exhibited a higher productivity at the 'dry' site, the site with minor or no groundwater supply to the root zone.

Similar results to those for primary production are found in litter production. The lower organic input at the site without groundwater contact results in a long-term deficit of organic matter in the soil. Additionally, organic and mineral input to the humus layer which was formerly supplied by flood sediments is lacking.

Even at the patch scale the time scale of the biological adaption process ranges from several months to a few years to observe changes in the physiological functioning and it takes decades to observe a shift in the composition of the large plant communities. It could be demonstrated that the young trees below an age of about 15 years could adapt

to the hydrological change by extending the root system while old trees suffered most as was reflected in reduced productivity.

4. SPATIAL VARIABILITY OF THE SOIL WATER BALANCE

Full information required the simulation of the soil moisture characteristics and was only available for several locations. To regionalize the soil water data different data sets including water table gauges, observations of the groundwater table, terrain data and soil maps were selected.

The surface water table was calculated with a 1-D hydraulic model utilizing the profile data of ditches and oxbow lakes. A terrain model was used to delineate the inundated area for different flood frequencies of return periods of between 1.5 and 12 years. Prior to the implementation of the scheme the 1.5-year flood caused local inundations covering a strip of 100 m along the channels and oxbow lakes. After the impounding in 1976 there were no inundations associated with the 1.5-year flood. Inundations caused by floods with higher return periods (over 15 years) will remain unchanged in areal extension. In any case the dynamics of flooding and the duration of inundation are drastically reduced.

Table 1. *Dry weight of biomass and detritus at two forest stands in the flood plain.*

	Biomass (g m^{-2} a^{-1})		Detritus (g m^{-2} a^{-1})	
	Dry site	Wet site	Dry site	Wet site
Tree layer	582	689	382	518
Shrub layer	70	98	121*	131*
Herb layer	7	5		

Note:

* sum from shrubs and small trees.

The simulation results outlined in the previous section indicate that the major part of the area and also the vegetation are strongly dependent on the fluctuations of the groundwater table. Only for shallow and sandy soil layers does the precipitation constitute the main driving hydrological input for the temporal variation of the soil moisture content in the upper soil layer.

To regionalize the effect of groundwater on the soil moisture balance a probabilistic approach was used. It can be concluded from the soil moisture simulations that the total number of days in which the groundwater table reaches the soil column is a good indicator for the soil moisture in the root zone.

Therefore, the effect of a change in the groundwater level on the soil water budget is expressed by the 'probability of groundwater contact' $P(h|x)$ for location x where the elevation is given. The thickness of the soil layer was obtained by drillings at 380 sites and was subsequently grouped with respect to the various soil types. Considering the terrain elevation a distribution function of the soil thickness $F(h)$ was empirically established for each soil type and forest stand unit. The probability of groundwater influence was estimated from

$$P(x) = \int P(h|x)F(h)dh \qquad (8)$$

as exhibited in Fig. 7. For a given location x the probability density function of the groundwater level can be obtained from the groundwater observations which have to be spatially interpolated, e.g. by kriging techniques. The terrain which is given in its elevation along cross profiles located at a distance of 500 m is modelled by a Voronoi net. The soil thickness represents the most critical spatial pattern because the spatial variability of the soil thickness even within a region of a specific soil type is remarkably high.

Considering the joint probability of groundwater influence (Fig. 7) as an indicator for soil moisture, a decrease of the groundwater fluctuation will result in a more pronounced heterogeneity in the spatial pattern of the hydrological conditions for the soil water balance . In the case of

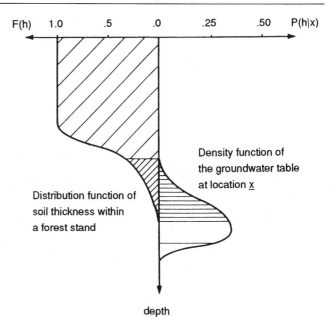

Fig. 7 Probability distribution function of soil thickness $F(h)$ and of the groundwater table depth $P(h|x)$ within a forest stand.

large groundwater dynamics all the subregions are affected by the groundwater table. With a decrease of the temporal variability the spatial pattern becomes more pronounced, some regions are always influenced by groundwater (wet sites) while others dry out.

A comparison of the 'probabilities of groundwater contact' before and after the dam construction shows at some locations the changes in the soil water budget induced by alteration of the groundwater regime (Fig. 8(a)).

5. SPATIAL VARIABILITY IN BIOLOGICALLY RELATED PARAMETERS

The change in the soil moisture balance (Fig. 8(b)) exhibits a quite heterogeneous spatial pattern. Obviously, the patchiness in the soil moisture distribution is increased. The biomass production of the vegetation promptly reflects the modified water supply while the response in the composition of the plant communities is remarkably delayed. Also, diffusion processes and competition among plant species partly compensate the hydrological tendency towards increasing patchiness. This will be demonstrated by a simplified example to obtain some information about the critical patch size.

Assume a site which exhibits throughout the year rather 'wet conditions' and which is surrounded by a quite different environment with 'dry conditions'. The diameter of the site is L and the growth rate of a plant species well adapted to a humid regime is R, while outside L the growth is zero (Fig. 9).

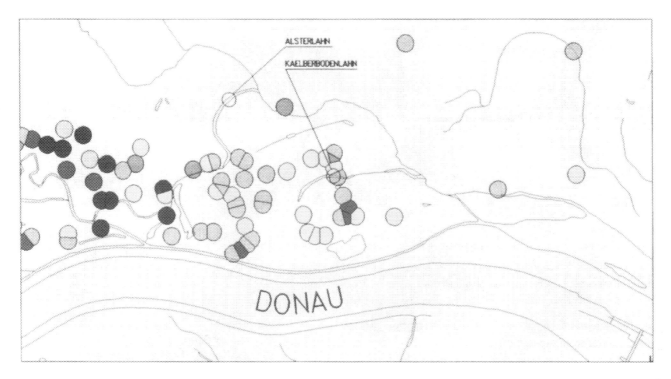

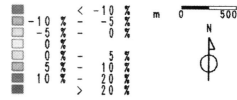

Fig. 8 (a) Changes in probability of groundwater supply to the soil column at several locations as induced by the construction of the Altenwörth power plant.

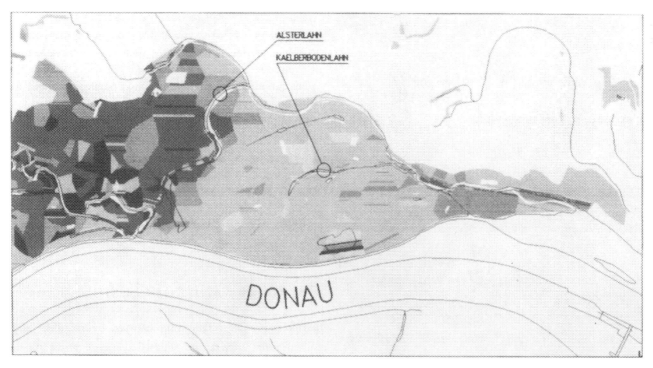

(b) Regionalized changes in the probability of groundwater supply to the soil column within the region (decrease of groundwater supply from west to east).

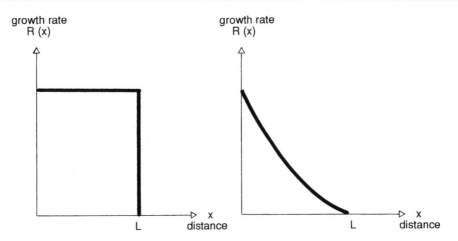

Fig. 9 Diagrams of simple growth rate models.

In the one-dimensional example the population $S(x,t)$ develops according to eqs. (9)–(12)

$$\frac{\delta S}{\delta t} = D\frac{\delta^2 S}{\delta x^2} + RS \qquad (9)$$

with D being the diffusion coefficient and R being the growth rate.

Applying as initial conditions and boundary conditions respectively

$$t = 0, \quad S = S_0 \qquad (10)$$

$$S = 0 \quad \text{for } x \le 0 \text{ and } x \ge L \qquad (11)$$

the solution of eq. (9) is

$$S(x,t) = \Sigma O_j \sin\left(\frac{j\pi x}{L}\right) \exp\left(R - j^2\pi^2\frac{D}{L^2}\right)t \qquad (12)$$

It can be easily concluded that for

$$L > \pi\left(\frac{D}{R}\right)^{1/2} \qquad (13)$$

the population increases in time while in the opposite case the diffusive losses cannot be compensated by the productivity and thus the species disappears. The critical patch size is

$$L_{cr} \ge \pi\left(\frac{D}{R}\right)^{1/2} \qquad (14)$$

As the species is in a hostile environment it will slowly diminish if the diameter of the patch is less than L_{cr}. If competition among plants is being considered explicitly a larger L_{cr} is required.

Some of the rigorous assumptions can be mitigated so that a spatially variable growth rate $R(x)$ is assumed.

Guney and Nisbet (1975) investigated the growth rate according to the following functions:

$$R(x) = r_0 - r_1 V(x) \qquad (15)$$

$$V(x) = \frac{r_0}{r_1}\left(\frac{x}{L}\right)^{-1} \qquad (16)$$

In this case eq. (12) is transformed into the Schrödinger equation (Schrödinger, 1928) and again the critical patch size is obtained from the set of eigen values.

Further, competition among species can be explicitly incorporated into the model. This implies that two equations similar to (9) have to be coupled. For all these cases, the principal conclusions obtained from eq. (13) still hold and a critical patch size will be obtained. Roughly speaking, the biological system averages in the long term the heterogeneous spatial parameters, such as soil moisture distribution.

The smoothed pattern of the hydrological changes as reflected by the biological system is given in Fig. 10.

It can be concluded that the flood plain forests which under natural conditions exhibit a pronounced patchy structure will under modified hydrological conditions approach a rather homogeneous spatial distribution. The main reason is the fact that flood events which interrupted the successional process are reduced in their frequency and magnitude. The successional development becomes dominant and averages the patchy structure.

6. SUMMARY AND CONCLUSIONS

In this paper the relationship between hydrological and biological parameters was analysed at several scales. As an example, the flood plain forest located along the Austrian section of the Danube was selected.

Due to the major changes of the hydrological boundary conditions caused by the implementation of impoundment dams, the states of both the hydrological regime and the biological system were modified. As one of the main indi-

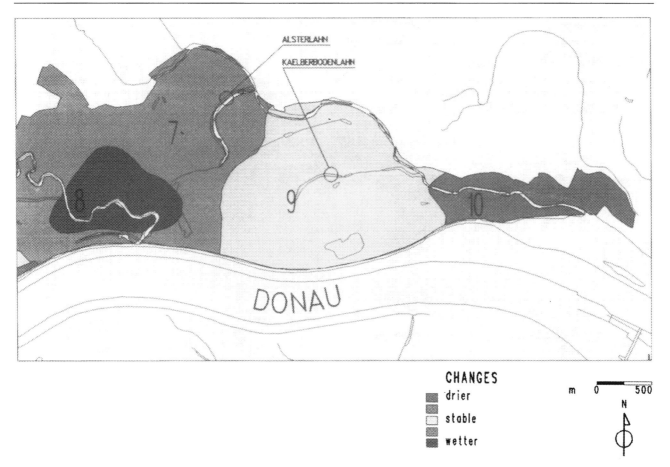

Fig. 10 Smoothed changes in the soil water balance (decrease of groundwater supply from west to east).

cators which link the hydrological system and the biology, the soil moisture balance was identified and analysed in its spatial and temporal pattern.

The soil moisture balance in the flood plain is dependent on inundation, the depth of the groundwater table, precipitation, climate and biology. After the impoundment the flooding frequency and the groundwater dynamics are remarkably reduced. Therefore the soil moisture balance becomes strongly dependent on the groundwater table which supplies the soil column. For selected sites the model simulations indicate that the water supply of the root zone is mainly dependent on the higher moments of the probability density function of the groundwater table while the mean value is of minor importance.

A change in the soil water balance is reflected differently by the vegetation layers. A decrease of the water supply results promptly in a decrease of the biomass production and of the LAI of the tree and shrub layer. The decrease in the LAI, the increased energy input and throughfall stimulate the productivity of the herb layer. Its contribution to total biomass production is small and cannot by far compensate the biomass losses in the other layers.

Extending the scope from the patch to the regional scale it was found that the decrease of the temporal variability in the

groundwater table results in an increased spatial variability of the soil moisture balance. Prior to the impoundment the broad range of fluctuations of the groundwater table supplied the soil layers in almost the whole region. After impoundment the variability in the terrain elevation and in the soil thickness is no longer smoothed out by the groundwater dynamics.

Therefore, the spatial heterogeneity in the soil water balance is enhanced. The small scale pattern in the hydrological parameters is not directly reflected in the vegetation. Although biomass production reacts quickly to the modified hydrological regime the plant communities adapt within a remarkable delay to the new hydrological conditions. A decrease in the hydrological temporal variability stimulates successional development which is no longer counter balanced by perturbations such as flood events. Growth rates, competition and diffusion processes determine a critical patch size for species requiring a specific hydrological regime. Small scale hydrological variability is in the long term averaged out by the plant communities and their successional development.

The considered time scales range from a few weeks for the hydrological parameters to one or two years when biomass production is considered. The ecosystem characterized by the plant communities in the flood plain changes much more slowly and time scales from several decades to hundreds of years should be envisaged.

ACKNOWLEDGEMENTS

This work was supported by the Austrian Ministry of Sciences and was carried out under contract of the Austrian Academy of Sciences. The Eco-system case study Altenwörth is an Austrian contribution to UNESCO's Man and Bioshere Program 5.9.

REFERENCES

CAMPBELL, G.S. (1985). Soil physics with basic transport models for soil-plant systems. Developments in Soil Science 14, Elsevier, Amsterdam.

CHANG, H.H. (1988). Fluvial processe in river engineering. pp. 432, J. Wiley and Sons, NY.

DECAMPS, H., M. FORTUNE, F. GAZELLE and G. PAUTOU (1988). Historical influence of man on the riparian dynamics of fluvial landscapes. Landscape Ecology, 1, 163–173.

FEDDES, R.A., P.J. KOWALIK and H. ZARADNY (1978). Simulation of field water use and crop yield. PUDOC, Wageningen, pp. 192.

FEDDES, R.A., P. KABAT, P.J.T. VAN BAKEL, J.J.B. BRONSWIJK and J. HALBERTSMA (1988). Modelling soil water dynamics in the unsaturated zone – State of the art. J. Hydrology 100, 69–111.

FRISELL, C.A., W.J. LISS, C.E. WARREN and M.D. HURLEY (1986). A hierarchical framework for stream habitat classification; viewing streams in a watershed context. Environmental Management, 10, 199–214.

GEPP, J. (1985). Water bodies in the Austrian flood plains. Summary and analyses of a reduced variety. Green Series of the Ministry of Health and Environmental Protection, Vol. 4, Vienna, Austria (in German).

GUNEY, W.S.C. and R.M. NISBET (1975). The regulation of inhomogeneous populations. J. Theoret. Biol. 32, 441–457.

HARY, N. and H.P. NACHTNEBEL (1989). Ecosystem study Altenwörth – Austrian contribution to UNESCO's 'Man and Biosphere Program', Vol. 14, pp. 443, Univ. Vlg. Wagner, Innsbruck, Austria (in German).

KIMMINS, J.P. (1987). Forest Ecology. McMillan Publ. Comp. NY.

LOCKWOOD, J.G. (1985). World Climatic Systems. Chapter 5 in: Grassland and Vegetated Subsystems. E. Arnold, Publ.

MAIER, R. (1989). The ecology of flood plain forest: physiological aspects of the impacts of different groundwater tables on the vegetation. In 'Ecosystem study Altenwörth'. Eds.: Hary, N. and H.P. Nachtnebel, vol. 14, pp. 443, Austrian contribution to UNESCO's 'Man and Biosphere Program'. Univ. Vlg. Wagner, Innsbruck, Austria (in German).

MARKAR, M.S. and R.G. MEIN (1987). Modelling of evaporation from homogeneous soils. Wat. Resour. Res. 23 (10), 2001–2007.

NACHTNEBEL H.P. and S. HAIDER (1991). Hydrological basis for the management of soil water budget in a riverine forest. In 'Hydrological Basis for Ecologically Sound Management of Soil and Groundwater', Eds.: Nachtnebel H.P. and K. Kovar, pp. 335–345, IAHS. Publ. No. 202, IAHS-Press, Wallingford, Oxon, UK.

PETTS, G.E. and H. MÖLLER AND A.L. ROUX (Eds.) (1989). Historical Change of Large Alluvial Rivers, Western Europe. J. Wiley, Chichester, UK.

PINAY, G., H. DECAMPS, E. CHAUVET and E. FUSTEC (1990). Functions of ecotones in fluvial systems. In The ecology and management of aquatic-terrestrial ecotones. Eds.: Naiman, R.J. and H. Decamps. Man and Biosphere Series, Vol. 4, UNESCO, Paris.

PRASAD, R. (1988). A linear root water uptake model. J. Hydrology 99, 297–306.

REILY, P.W. and W.G. JOHNSON (1982). The effect of altered hydrologic regime on tree growth along the Missouri River in North Dakota. Can. J. Bot. 60, 2410–2423.

SCHRÖDINGER E. (1928). Wave mechanics. Blackie and Sons, London, UK.

Problems and progress in macroscale hydrological modelling

A. BECKER

Potsdam Institute for Climate Impact Research,

Telegrafenberg, PF 601203, D-14412 Potsdam,

Germany

ABSTRACT One of the major problems in macroscale hydrological modelling is the assessment of areal heterogeneity in important land surface characteristics, such as topography, land use, land cover, soil, vegetation and hydrological characteristics. After a brief discussion of spatial scales to be covered and related categories of models to be applied two examples are presented which underline the problems involved in the application of lumped models for large areas, like grid areas of general atmospheric circulation models (GCMs). A strategy for a more appropriate hydrologically sound structuring of macroscale hydrological models is then outlined which takes into account the following facts and features: (1) zones of 'uniform' atmospheric forcing, (2) landscape patchiness, (3) intra-patch heterogeneity, (4) a 'Two-Domains-Modelling' concept, which is essential for the coupling of atmospheric and land-surface hydrological models.

1. INTRODUCTION

Global modelling is the subject of two of the most challenging recent international programmes:

- The World Climate Research Programme (WCRP), and
- The International Geosphere-Biosphere Programme (IGBP) – a Study of Global Change.

A main objective in both programmes is to improve land-surface process descriptions (parameterisations) to be applied at large scales, up to the scale of grid areas of global atmospheric circulation models (GCMs) which cover about 10^4–10^5 km^2.

Problems in large scale land-surface process modelling and some improvements achieved during the last years are briefly discussed in the following and suggestions are made for further progress.

In principle it is understood that a hypothetical planet 'Modellion', as recently defined by Shuttleworth (1991), is needed to understand possible future developments better, and to identify the action necessary to avoid or reduce negative consequences for the welfare of the humanity.

2. SPACE SCALES TO BE COVERED

In the left part of Table 1 a definition of scales is given which takes into account the space ranges of the three main categories in scaling (micro-, meso-, macroscale), and also two 'transition ranges' (according to meso-α and meso-γ in meteorology):

about 10^3 to 10^4 km^2 – transition range meso-/macroscale (30 to 100 km)

about 10^{-4} to 10 km^2 – transition range micro-/mesoscale (10 m to 3 km)

3. DIFFERENT MODEL CATEGORIES AT DIFFERENT SCALES

It has been found that in land-surface hydrology different modelling principles and approaches need to be applied in different scales. In the microscale the application of elementary process models, e.g. in the form of fundamental differential equations of hydro- and thermodynamics, is most appropriate for single leaves, single plants, soil columns and pedons.

Table 1. *Scales, related subjects and areas in modelling.*

Scales and sub-scales	Related areas in modelling
MACROSCALE > 100 km (> 10^4 km^2)	Global atmospheric circulations models (GCMs)
MESOSCALE-α $\sim$30–10^3 km ($\sim 10^3$–10^6 km^2)	Regional climate, single GCM grid
MESOSCALE-β $\sim$3–30 km (10–10^3 km^2)	Heterogeneous landscapes, biomes Complex river basins Planetary boundary layer (PBL)
MESOSCALE-γ $\sim$1–5 km ($\sim$1–25 km^2)	
MICROSCALE $\leq$10 m ($\leq$100 m^2)	Patches, ecotopes Single plants, soil columns, pedons Single leaves

These equations are, of course, equally valid for large areas. At land surfaces, however, unlike the atmosphere and large water bodies, various heterogeneities, discontinuities, phase changes, and other changes occur so that the conditions of continuity and internal homogeneity are often not fulfilled. Therefore the applicability of the aforementioned fundamental laws, at least in a unique form, becomes questionable.

Dooge (1985) has formulated in this regard: Large scale hydrological modelling usually means that '. . . the finer scale processes may be either ignored or may be represented by their statistical effects in the large scale description' . . . 'It has been found in practice that the models based on continuum mechanics are too complex to allow for the spatially variable nature of hydrologic systems to be taken into account in large scale modelling and they have been simplified to such an extent that they became in effect simple conceptual models.'

These statements are essential in two respects:

(1) they confirm the statement made above on a direct application of the fundamental laws of hydro- and thermodynamics (category-1 models) for modelling hydrological processes at large land-surface areas;
(2) a different category of models, called 'conceptual' (category-2 models), is considered as more appropriate for those applications (meso- and macroscale).

In Fig. 1 an overview is given of the different categories of hydrological models, including the two mentioned above, and also of the most important spatial discretization schemes (Becker and Serban, 1990).

There are several well developed and widely applied conceptual hydrological models which can be classified as category 2 (e.g. Becker and Serban, 1990). Some of them are

physically-based in the sense that they were derived with acceptable simplifications from the equations mentioned in (1), or they represent appropriate analogues of physically sound models, for instance diffusion type models, storage reservoirs, translation elements, cascades of such elements etc.

Experience in applying these models indicates that they are most suitable for areas ranging from about 10 km^2 up to a thousand or a few thousand km^2 (about 3 × 3 up to 30 × 30 km), i.e. for heterogeneous landscapes, biomes, river basin areas, other land-surface units of an appropriate size, which significantly influence the processes in the planetary boundary layer (PBL). According to Table 1 these areas clearly belong to the mesoscale. Considerable losses in simulation quality occurred, however, when the same models and approaches were applied either to smaller or to larger areas.

Accordingly it may be concluded that:

(i) Hydrological land-surface process modelling in the microscale can be based directly on the fundamental differential equations of hydro- and thermodynamics with a detailed spatial resolution.
(ii) Conceptual physically-based models (category 2) seem to be more appropriate to the mesoscale.
(iii) In the macroscale range the model area may be subdivided into subareas of the mesoscale size and category-2 models may then be applied to these subareas.

Special comments on the application of category-2 models will be made later.

4. CRITICAL HYDROLOGICAL CONSEQUENCES OF RECENT PRACTICES APPLIED IN MACROSCALE MODELLING AT LAND SURFACES

From a hydrological point of view the application of the lumped modelling approach (see Fig. 1) in large scale hydrological modelling with averaging of the process characteristics (forcing inputs, state conditions, land surface properties, etc.) over large areas, as e.g. GCM grid areas of 10^4 to 10^5 km^2, is considered as very critical. This can be illustrated by means of the following two examples.

EXAMPLE 1

A GCM grid square covers predominately flat and hilly terrain (70% of area) where the long-term average annual precipitation *P* ranges from 300 to 500 mm (areal average 400 mm) and evapotranspiration *ET* from 280 to 420 mm (areal average 350 mm). At one side of the grid a mountain range is located (30% of the grid area) where *P* is significantly higher and ranges up to 1500 mm and, due to the availability of more water, *ET* up to 530 mm. The related areal averages are

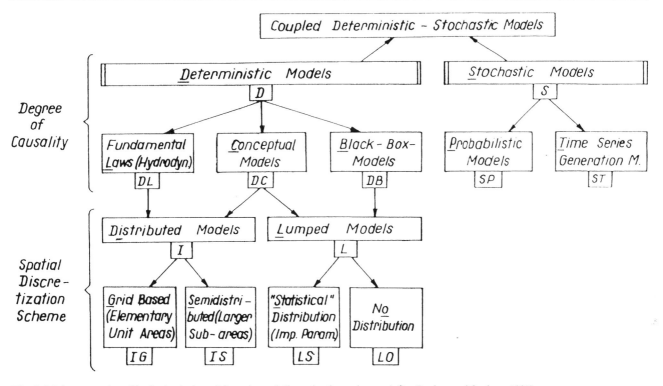

Fig. 1 Main categories of hydrological models and areal discretization schemes (after Becker and Serban, 1990).

listed in the following overview, together with the long-term areal average runoff ($R = P - ET$):

Flux (mm/year)	Lowland part (= 70% of the area)	Mountain part (= 30% of the area)	Whole grid area (= 100% of the area) (= 0.7 flux lowland area + 0.3 flux mountain area)
P	400	1000	580
ET	350	460	383
R	50	540	197

As can be seen the 30% mountainous area of the GCM grid produces 0.3×540 mm, i.e. nearly 83% of the total runoff from the grid area, while the larger lowland part produces only 0.7×50 mm, i.e. about 17%. It is evident that precipitation, as the primary forcing input into the system, is responsible for the significant difference in long-term average runoff between the mountainous part and the lowland part of the grid unit, and accordingly for the hydrological behaviour in general. If this difference is not taken into account explicitly, for instance in separate models for the two subareas, the physical soundness of the process description is automatically lost and parameterisations become 'arbitrary', even if the model has proved to 'fit well' in the 'calibration' period.

EXAMPLE 2

Another critical case is that of oases within a desert. The oases, including the surrounding vegetated area in contact with groundwater, may cover 2% of the reference area (e.g. a

GCM grid square). For simplification it may be assumed that no transition zone exists between the oases and the desert which varies in size depending on groundwater levels. The following long-term averages of evapotranspiration amount may be given:

Evapo- transpiration (mm/year)	Oases (= 2% of the area)	Desert (= 98% of the area)	Whole grid area (= 100% of the area) (= 0.02 × 2600 + 0.98 × 15)
ET	2600	15	67

Evapotranspiration from the oases is identical with the potential rate. Although the partial area is only 2% of the reference area, it contributes about 78% ($0.02 \times 2600 = 52$ mm) to the total grid square evapotranspiration. The remaining 15 mm (about 22%) originates from evaporation during a few rainfall events in the desert area. The separate determination of ET for the two types of sub-areas is conceptually clear, physically sound and relatively simple, and so is the areal integration over the total GCM grid unit.

Any lumped model for the grid unit as a whole, however, 'violates' reality because it tries to simulate a desert area 'a bit wetter', so that it can produce an increased ET which in fact originates from the oases. Hydrological soundness is clearly lost and model parameters become just 'tuning variables.'

Therefore the application of lumped models for large areas like GCM grid units as a whole can only be tolerated until more suitable distributed or semi-distributed models can be applied, and as long as other shortcomings in GCMs are more critical.

Another critical approach in large scale hydrological land-surface modelling is the application of microscale models at large scales up to the macroscale of GCM grid units. Most of the 'soil-vegetation-atmosphere-transfer schemes' (SVATs) used in the past for describing land-surface processes in GCMs are 'big leaf – big stoma' representations and lumped. They imply that land is homogeneously covered by a single big leaf within a grid element of the numerical climate model. This big leaf usually has a single stoma which is sensitive (in the most sophisticated parameterisations) to the environmental conditions known to have an effect on the mechanism of the stomata (i.e., solar radiation, temperature, humidity, carbon dioxide and soil water pressure head in the root zone). This stoma controls the plant transpiration and, as a result, the Bowen ratio and the surface heat fluxes over the GCM grid element as a whole (Avissar, 1991).

It is evident, in view of the natural landscape variability and the non-linear nature of the involved processes, that this approach is conceptually inappropriate and needs to be improved. Among others Fiering (1982) has commented on this problems as follows: 'One of the assumptions frequently made is that our understanding of the microscale elements and processes (in the hydrological cycle) can, with minor modifications, be extrapolated in principle to the understanding of the macroscale environment, thus enabling reliable predictions to be made by linking the solutions to form a causal chain. Unfortunately, it seldom happens that way. Sooner or later, at some scale or characteristic dimension, mechanistic explanation breaks down and is necessarily replaced by unverified causal hypotheses or statistical representations of the processes.'

A large number of existing SVATs have recently been reviewed by Geyer and Jarvis (1991). This review includes comments on the scales of application.

The limitations of infiltration models from the point of view of large scale application have been examined by Entekhabi and Eagleson (1991) who developed an improved land-surface parameterisation scheme, which takes into account in a statistical way the areal distributions of precipitation and important land-surface characteristics.

In summary it must be said that it is time to overcome the shortcomings in macroscale hydrological modelling as characterised above.

5. HYDROLOGICALLY SOUND STRUCTURING OF MACROSCALE MODELS

It was in particular Klemes (1985) who critically examined the transportability of existing hydrological models and their applicability for long-term predictions in different environments and for changing conditions. He concluded that models for predictions should fulfil the following general requirements:

(1) the model structure must have a sound physical foundation and each of the structural components must permit its separate validation;
(2) the models must be geographically transferable, and their parameters must be derivable from estimates of real-world characteristics at any place on the Earth.

These requirements are in agreement with what has been said above. The question is now what possibilities are there for adequate, physically sound vertical and horizontal structuring of macroscale hydrological models.

Fig. 2 represents a schematic representation of the land-surface related subprocesses of the hydrological cycle and subsystems to be considered in hydrological modelling. It indicates which processes belong to the domain of vertical energy and moisture fluxes at land surfaces ('*Vertical Fluxes Domain*') and which to the domain of lateral surface and subsurface flows ('*Lateral Flows Domain*'). The two domains are sometimes classified by the terms 'water balance' model and 'water transport' model, respectively, e.g. by Vörösmarty and Moore (1991).

Modelling of the vertical fluxes can be related to any land-surface area of interest (grid units, patches, river basins etc.), while water transport models must principally take into account the boundaries of hydrological systems like river basins, groundwater systems etc., i.e. the drainage basin water divides. Considering this Becker and Nemec (1987) have suggested a so-called 'two-level' modelling concept in order to ensure that different areal discretisation schemes can be applied to the two 'levels' which were in fact understood as the two domains defined previously. In the following the term 'two-domains modelling concept' is used.

The following facts are particularly essential for the distinction between the two domains in modelling:

(i) In the *vertical fluxes domain* any required or useful areal discretization scheme can be applied, in a more or less detailed form:

   – a regular grid of any size (GCM grid or sub-grids),
   – a subdivision of the reference area into subareas of 'equal' or uniform hydrological behaviour, with respect to land surface properties, forcing atmospheric inputs, process characteristics, etc.

   An a priori relationship to watersheds (river basin divides etc.) is not necessary.

(ii) Modelling in the *lateral flows domain*, however, must be related from the very beginning to watersheds, even in large scale regional and continental modelling. Component models for this domain are clearly watershed related. In many cases simplified physically based or conceptual

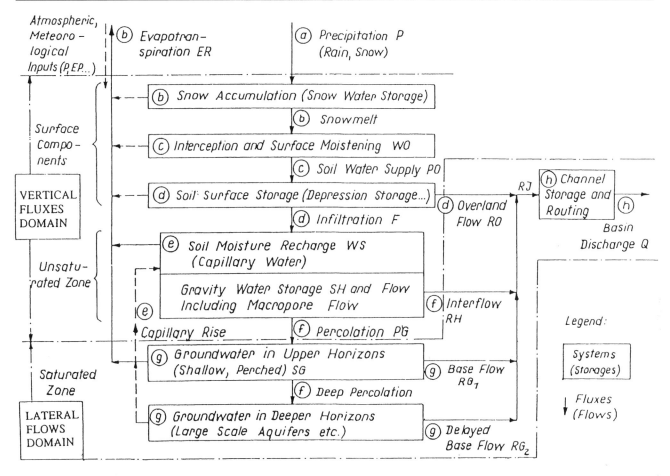

Fig. 2 Components and subprocesses of the land phase of the hydrological cycle.

models are sufficient to be applied to river basins or subbasins as a whole (category-2 models).

(iii) In river basin studies only those parts of the vertical fluxes domain model which lie within the river basin need to be coupled to the lateral flows domain.

For GCMs the vertical fluxes domain is particularly important. Lateral flows domain models are of interest only insofar as observed flows are intended to be used for validation purposes, or for the investigation of interactions between water resources systems and climate.

The essential question in modelling the vertical fluxes domain is what areal discretisation scheme should be applied to guarantee a hydrologically sound assessment of areal heterogeneity and to avoid problems as discussed in the previous section. The main alternatives are represented in Fig. 1 (Becker and Serban, 1990). The application of a high resolution fine meshed regular grid is always possible but troublesome and not always efficient. The lumped approach is very critical. Therefore the semi-distributed approach is considered as an appropriate compromise solution. It allows one to take into account, in an appropriate way, areal process differentiation as discussed before. Subareas to be modelled separately can be chosen as large as possible. They don't need to be divided into smaller pieces.

6. BASIC TYPES OF AREAL HETEROGENEITY AT LAND SURFACES AND ASSESSMENT OF LANDSCAPE PATCHINESS

There are two basic types of heterogeneity that need to be considered (Avissar, 1991): (a) landscape patchiness and (b) intra-patch heterogeneity.

Landscape patchiness means the succession (mosaic) of different types of land surface area: agricultural fields, forests, built-up areas, water surfaces etc. on various topographical, pedological, geological and other features. Patches generally represent hydrological units with an 'approximately equal' or 'uniform' hydrological behaviour.

Intra-patch heterogeneity refers to the small scale variability of soil, vegetation and other parameters, for instance in a vegetated patch, rooting depth, stomatal resistance, etc. Intra-patch heterogeneity is generally more or less random in its distribution so that statistical approaches are best suited for its assessment. Patches with clearly distinct hydrological regimes may require model components that are differently structured.

The first important question in assessing landscape patchi-

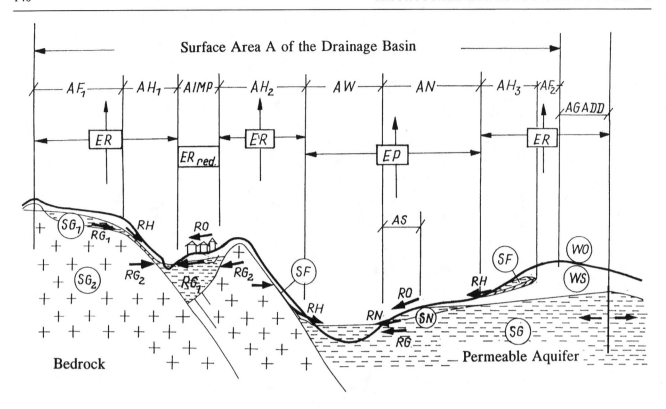

Fig. 3 Cross-section of a complex landscape (river basin) with indication of subareas of different hydrological regime ('patch types').

ness is how many types of patch need to be taken into account.

Penman who looked at the problem from the point of view of evapotranspiration suggested in 1951 two classes of land surface should be distinguished:

– Riparian areas where the groundwater table would be close enough to the surface for evaporation to be always at the potential rate (shallow groundwater areas with a capacity for exfiltration).

– Non-riparian areas where groundwater would be well below the surface and the actual rate of evapotranspiration during dry periods would be solely dependent on the moisture content in the unsaturated zone (areas with deep groundwater with a high capacity for infiltration).

This is a very essential approach which immediately solves the 'oasis problem' as stated before, as well as many similar problems in other landscapes.

Subsequently, when runoff formation is taken into account, hillslope areas and also built-up areas with a high percentage of impervious surface have to be additionally separated to improve the simulation of direct runoff (overland flow and interflow). A review of this development has been given by Kuhnel et al. (1991). A schematic representation of several land-surface types, differing in important hydrological process characteristics, is given in Fig. 3.

The classification of the patch types presented in Table 2 (Becker and Pfützner, 1986) is based on Fig. 3. It considers

both evapotranspiration and runoff formation and represents an appropriate compromise towards a hydrologically sound discretization of land-surface areas. Of course, other patch types (forests, pasture, different crop stands) can additionally be taken into account, if this is essential.

In any case, the decision of how many types of patches need to be considered should be made carefully. It depends strongly on the problem to be solved. In macroscale modelling it will often be sufficient to distinguish between only two evapotranspiration related types, $(AW + AN)$ in Table 2 on one hand, and all other types on the other. More appropriate may be to consider four types: AW, $AIMP$, $(AF + AG + AH)$, AN.

When dealing with the components of runoff, the six types of Table 2 should be distinguished. For more detailed bisophere studies, consideration of different crop stands may also be required.

After it has been decided which patch types to take into account the subdivision (disaggregation) of the model area into subareas can be done (on the basis of maps, air photographs etc.). An important simplification can then be achieved by combining all single patches of a special type within the model area into one arbitrary unit. This unit can then be modelled with an adequate component model.

If the accumulated total area of a special patch type i within a model area (e.g. a grid element) is defined as A_i then the total flux of a certain type k (sensible heat, latent heat,

Table 2. *Patch types with significantly different hydrological regimes (after Becker and Pfützen, 1986) being based on Fig. 3.*

Patch Type	Symbol	Evapotranspiration	Direct Runoff
Water areas	AW	Potential	Equal to precipitation (100%)
Impervious areas (sealed, rocks)	AIMP	Only during precipitation events (i.e. generally small)	Nearly equal to precipitation (near to 100%)
Flat areas with:			
– Deep groundwater (infiltration areas)	AF	Depending on the soil moisture in the root zone	Generally near to zero
– Shallow groundwater (exfiltration areas)	AN	Generally potential	With increasing water storage equal to precipitation at saturated subareas
Hillslopes with:			
– Deep permeable underground (infiltration areas)	AG	As AF above	As AF above, but infiltration excess possible during heavy rainfall or with frozen ground
– Shallow soils above less permeable layers or bedrock	AH	As AF above	From saturated subareas equal to precipitation otherwise infiltration excess and interflow

long-wave radiation or others) from the whole model area is given as

$$F^{(k)} = \frac{\sum_{i=1}^{n} A_i f_i^{(k)}}{\sum_{i=1}^{n} A_i} \tag{1}$$

with $f^{(k)i}$ being the particular flux of type k from patch type i, and n the number of patch types considered.

This approach clearly represents a '*semi-distributed*' approach according to Knudsen et al. (1986) and Becker and Serban (1990). It neglects the real distribution of the different individual patches within the model area and their interaction with neighbouring patches of different types (advection effects etc.), but it overcomes the critical effects of averaging the different hydrological behaviours of different patch types over large areas.

There are several examples of the successful application of this modelling technique for river basin modelling (e.g. Knudsen et al., 1986; Becker and Pfützner, 1986; Avissar and Pielke, 1989; Kaden and Pfützner, 1991). They cannot be described in the framework of this paper. However, in summary it can be said that the appropriateness, usefulness and efficiency of the semi-distributed approach for large scale modelling has clearly been confirmed.

As has already been stated, intra-patch heterogeneity is usually random and occurs even on small scales (microscale). Accordingly it is more appropriate to assess this heterogeneity by means of statistical approaches, e.g. by applying distribution functions for the relevant land-surface and other characteristics in a model of the study area such as a patch.

Several such approaches have been developed and are described elsewhere. They concern the following processes and properties:

(a) water holding capacity of the rooted soil layer (Becker, 1975);
(b) runoff ratio, soil evaporation and transpiration (Entekhabi and Eagleson, 1989);
(c) different soil and plant characteristics (Avissar, 1991);
(d) topography, soil and plant characteristics (Famiglietti and Wood, 1991).

All these approaches are steps in the desired direction. However, they are generally applied without considering the constraint that their application is limited to areas (zones) of uniform atmospheric forcing, as discussed earlier in example 1. In any case, further investigation is needed.

7. HIERARCHY IN THE AREAL DISCRETISATION OF LAND SURFACES FOR MODELLING

In conclusion a hierarchical approach is recommended in discretising (disaggregating) land surfaces for hydrological modelling.

(1) Investigation of the areal pattern of essential characteristics (for instance long-term mean) of atmospheric driving forces operative on hydrological processes at land surfaces; in particular precipitation, radiation and related characteristics.

(2) Definition of ranges of these characteristics within which areal averaging is acceptable, at least in the long-term (see

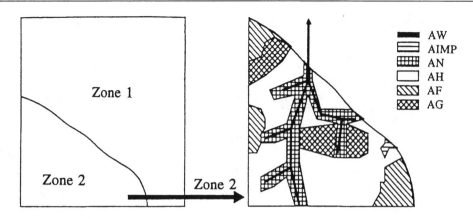

Fig. 4 Examples of areal disaggregation of a large land-surface area, such as a GCM grid area, into zones 1 and 2 of 'uniform atmospheric forcing' (left) and of zone 2 into subareas of different 'patch type' according to Table 2 (right).

Section 8), and statistical distribution functions can be used to describe event specific areal differentiations (patterns). This has been done for instance by Entekhabi and Eagleson (1991).

(3) Subdividing (zoning) of the study area into subareas (zones) of 'uniform' atmospheric forcing (in terms of long-term characteristics, in particular long-term means).

Remark: This zoning is essential and is needed to define larger areas which, from the point of view of atmospheric forcing, can be handled as units in large scale modelling. It avoids the problems explained earlier in example 1. Additional explanation will be given in Section 8. An example of the zoning is given in Fig. 4, left side (developed from Kaden and Pfützner, 1991).

(4) Definition of patch types to be modelled separately (considering the remarks made in Section 6).

(5) Assessment of landscape patchiness, i.e. subdivision of the study area into patches of different types and determination of the areal percentage of each patch type considered in each zone of uniform atmospheric forcing (aggregation of individual patches). An example is given at the right hand side of Fig. 4.

(6) Selection of an appropriate patch type related model which may (or may not) include elements for assessing intra-patch heterogeneity.

(7) Estimation of the parameters of the selected patch type related models.

After considering the approach described under (1)–(7), an integrated model for the 'vertical fluxes domain' of the study area can be composed by aggregating the component models of the subsystems involved, or at least the fluxes which they produce, according to eq. (1).

The vertical fluxes domain model allows one to describe and study all important hydrological processes at the land-atmosphere interface. The coupling of models to it for the lateral flows domain is a separate modelling step related to the aquifer and river basin, as has been discussed above. Modelling of this *lateral domain* is required if:

- the hydrological cycle needs to be closed downstream for a river basin, or as far as the ocean;
- vertical fluxes domain models are intended to be validated with reference to measured discharges;
- water resources impact studies are planned;
- interactions between climate and water resources are to be investigated;
- problems of material transport in rivers, of nutrient cycling etc. are to be solved.

It should be pointed out again that the two-domain-modelling concept allows one to model separately both domains with different areal discretization schemes and to couple both models finally with reference to the river basin divides (Becker and Nemec, 1987).

8. CRITERIA TO DELINEATE ZONES OF UNIFORM CLIMATE FORCING

As remarked upon in the previous section, the zoning of large land-surface areas (e.g. GCM grid units) into 'zones of uniform atmospheric forcing' is of particular significance. However, no general methodology is available for this purpose. Therefore, a proposal is made here which could contribute to such a methodology.

Let the local state or flux variable at instant t at any point x, y in a study area be $u(x, y, t)$. The long-term total of this variable is:

$$\bar{u}(x, y) = \int_{t_0}^{t_0 + T} u(x, y, t) dt \qquad (2)$$

where T is the integration period, e.g. 1 year or several years. A zone of uniform climate forcing ACF may then be defined as an area where the local amounts of $\bar{u}$ do not deviate significantly from the areal average $\bar{U}_{ACF}$:

$$\bar{U}_{ACF} = \int\int_{ACF} \bar{u}(x,y)dxdy \qquad (3)$$

If a tolerance deviation Δu is defined the following 'non-exceedance criterion' can be formulated to define a zone ACF:

$$|\bar{u}(x,y) - \bar{U}_{ACF}| < \Delta u \qquad (4)$$

Δu may be defined as a percentage of $\bar{U}_{ACF}$, for instance 20%.

The above criterion is related to long-term areal totals or means. There may be other criteria which can equally be used, for instance, variances, differences in seasonal variation patterns etc.

However, precipitation is particularly essential for the hydrology, the natural resources and the environment in particular areas or in a river basin in general. Long-term totals or averages of precipitation and related parameters determine the water balance in these areas. Therefore, the above criterion will be important. However, further studies in this field are indispensable.

REFERENCES

AVISSAR, R. (1991): A Statistical-Dynamical Approach to Parameterize Subgrid-Scale Land-Surface Heterogeneity in Climate Models: Surveys in Geophysics 12, 155–178.

AVISSAR, R. and PIELKE, R.A. (1989): A Parameterization of Heterogeneous Land Surface for Atmospheric Numerical Models and Its Impact on Regional Meteorology, Mon. Wea. Rev. 117, 2113–2136.

BECKER, A. (1975): The Integrated Hydrological Catchment Model EGMO. – In: Int. Symp. and Workshops Appl. Math. Models in Hydrology and Water Res.Syst., Bratislava. Hydrol. Sci. Bull. 21, 1.

BECKER, A. and NEMEC, J. (1987): Macroscale Hydrologic Models in Support to Climate Research. In: The Influence of Climate Change and Climatic Variability on the Hydrologic Regime and Water Resources. Proceedings of the Vancouver Symposium, August 1987. IAHS Publication No. 168, 431–445.

BECKER, A. and PFÜTZNER, B. (1986): Identification and Modelling of River Flow Reductions caused by Evapotranspiration Losses from Shallow Groundwater Areas. 2nd Scientific Assembly of the IAHS (Symposium S2, Budapest, July 1986). IAHS Publication No. 156, 301–311.

BECKER, A. and SERBAN, P. (1990): Hydrological Models for Water-Resources Systems Design and Operation. Operational Hydrology Report No. 34, WMO-No. 740, Geneva.

DOOGE, J.C. (1985): Hydrological Modelling and the Parametric Formulation of Hydrological Processes on a Large Scale, WCP-Publ. Ser. No. 96, WMO/TD-No. 43, Geneva.

ENTEKHABI, D. and EAGLESON, P.S. (1989): Land Surface Hydrology Parameterization for Atmospheric General Circulation Models Including Subgrid Scale Spatial Variability. J. Climate 2(8), 816–831.

ENTEKHABI, D. and EAGLESON, P. (1991): Climate and the Equilibrium State of Land Surface Hydrology Parameterizations. Surveys in Geophysics 12, 205–220.

FAMIGLIETTI, J.S. and WOOD, E.F. (1991): Evapotransipation and Runoff from Large Land Areas: Land Surface Hydrology for Atmospheric General Circulation Models. Surveys in Geophysics 12, 179–204.

FIERING, M. (1982): Overview and Recommendations. In: Scientific Basis of Water Resources Management, National Academy Press, Washington, DC.

GEYER, B. and JARVIS, P. (1991): A Review of Models of Soil-Vegetation-Atmosphere-Transfer-Systems (SVATS). TIGER III, University of Edinburgh, EH9 3JU, March 1991.

KADEN, S., and PFÜTZNER, B. (1991): Large-Scale Modelling of Land Surface-Atmosphere Interface Processes – Macrohydrologic Software System. Report to UNESCO – IHP, Paris (SC/RP 205084,1).

KLEMES, V. (1985): Sensitivity of Water Resource Systems to Climate Variations. WCP Report No. 98, WMO, Geneva.

KNUDSEN, J., THOMSEN, A. and REFSGAARD, J.C. (1986): WATBAL – a Semi-Distributed, Physically Based Hydrological Modelling System. Nordic Hydrology, Vol. 17, No. 415.

KUHNEL, V., DOOGE, J.C.I., O'KANE, J.P.I. and ROMANOW-ICZ, R.J. (1991): Partial Analysis Applied to Scale Problems in Surface Moisture Fluxes. Surveys in Geophysics 12, 221–247.

SHUTTLEWORTH, W.J. (1991): Modellion Concept. Rev. Geophysics, 29, 4 (Nov. 1991), 585–606.

VÖRÖSMARTY, C.J. and MOORE, B. III (1991): Modelling Basin-Scale Hydrology in Support of Physical Climate and Global Biogeochemical Studies: An Example Using the Zambezi River. Surveys in Geophysics 12, 271–311.

Predictability of the atmosphere and climate: towards a dynamical view

C. NICOLIS

Royal Meteorological Institute of Belgium,

3, Av. Circulaire, 1180 Brussels,

Belgium

ABSTRACT Geophysical phenomena are often characterized by complex, random-looking deviations of the relevant variables from their average values. Traditional approaches attribute this complexity to external uncontrollable factors and to poorly known parameters, whose presence tends to blur some fundamental underlying regularities. In this paper we consider that complexity might be an intrinsic property generated by the nonlinear character of the system's dynamics. Bifurcations, chaos and fractals, three important mechanisms leading to complex behaviour in nonlinear dynamical systems, are reviewed and the role of the theory of nonlinear dynamical systems as a major tool of interdisciplinary research in geosciences is stressed. The general ideas are illustrated on the dynamics of fluctuation-induced transitions between multiple climatic states with special emphasis on Quaternary glaciations.

1. INTRODUCTION

Much of our understanding of the earth's past environmental and climatic conditions rests on the ability to decipher geological or atmospheric data. Now, a typical *time series* obtained from such data displays considerable complexity, reflected by the lack of any obvious periodicity and by the occurrence of random-looking excursions of the relevant variables from their average level. A question of obvious concern is, therefore, how to decipher the message of such a time series and how to attribute to the systematic effects and to the randomness the roles that they actually deserve.

When confronted with a complicated, or even at first sight erratic succession of events the simplest explanation that comes to mind is that the phenomenon of interest is blurred by the presence of a great number of parasitic variables and poorly known parameters which hide some fundamentally simple underlying regularities. Traditional statistical methods provide useful algorithms for extracting the relevant signal from what is believed to be random background noise. Our principal aim in the present paper is to draw attention to the fact that, in addition to such considerations, it is important to adopt complementary approaches as well,

based on a more dynamical view of the underlying phenomena (Nicolis and Nicolis, 1987).

The need for a dynamical approach stems from recent developments in physical and mathematical sciences indicating that the stable and reproducible motions that have dominated science for centuries no longer symbolize our physical world (Nicolis and Prigogine, 1989). Experiments on quite ordinary physical systems at the laboratory scale and the study of mathematical models reveal the existence of instabilities leading to the amplification of small effects and driving the systems to alternative states. These *nonlinear phenomena* are sources of intrinsically generated complexity and unpredictability. Furthermore some of the states arising through this mechanism present the characteristics of *deterministic chaos*: despite the fact that they are generated by a well-defined set of laws they give rise to an aperiodic, random-looking behaviour. Clearly, in the light of the above, in large classes of natural phenomena it may be meaningless to eliminate the variability and to keep only the mean as being the most representative part of the behaviour.

In the next section we summarize the principal mechanisms leading to complex behaviour in nonlinear dynamical systems. As an illustration of these ideas we subsequently

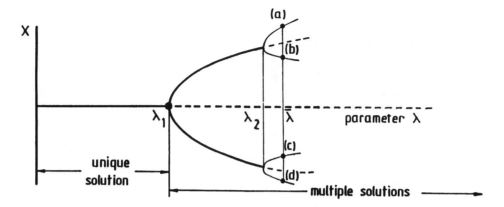

Fig. 1 Typical bifurcation diagram of a nonlinear dynamical system.

consider the dynamics of Quaternary glaciations in connection with the oxygen isotope record of deep sea cores. Finally we close with comments on the potential interest of the approach developed in this paper.

2. NONLINEAR DYNAMICAL SYSTEMS: BIFURCATION, CHAOS AND FRACTALS —

The evolution of a natural system in the course of time is conditioned by two major factors: a set of laws governing the individual elements constituting the system and their interactions, and a set of constraints acting from the external world, which are usually manifested via a number of *control parameters*. Typical examples are the solar constant, the concentration of trace gases, and so forth.

Let X_i, $i = 1, \ldots, n$, be the relevant variables. In accordance with the above considerations we write their rate of change in time in the form

$$\frac{dX_i}{dt} = v_i(X_1, \ldots, X_n; \lambda, \mu, \ldots) \quad i = 1, \ldots, n \quad (1)$$

where v_i stand for the evolution laws and λ, $\mu, \ldots$ for the control parameters. An illustration of eq. (1) is provided by the primitive equations describing atmospheric circulation or by the coupled energy and mass transfer between the hydrosphere, cryosphere and atmosphere.

Whatever the detailed interpretation of eq. (1) might be, a common feature shared by large classes of systems is that v_i are *nonlinear* functions of the state variables. This ubiquity of nonlinearity in nature stems primarily from the numerous *feedbacks* exerted between the different components of the system (surface-albedo feedback, etc.) Additional sources of nonlinearity may arise, for example from the presence of the Reynolds stress tensor in hydrodynamic flow.

Nonlinearity is a source of intrinsically generated complex behaviour and unpredictability, in the sense that more than

one outcome of the evolution now becomes possible. Fig. 1 depicts a typical scenario of the way the solutions X of a nonlinear dynamical system behave when a parameter built in it is varied. At the values $\lambda_1, \lambda_2, \lambda_3, \ldots$ of the parameter, referred to as *bifurcation points*, new branches of solution are generated. In general these bifurcation cascades produce a multiplicity of simultaneously available states (branches (a) to (d) for the value $\lambda = \bar{\lambda}$ in Fig. 1). Which of these states will actually be chosen depends on the initial conditions. This property confers upon the system a high sensitivity and a limited predictability, since the initial conditions are history-dependent and may be modified by fluctuations or by external perturbations. In actual fact therefore the dynamics of a multistable system will be an aperiodic succession of intermittent jumps between coexisting attractors. This view, which will be taken up in more detail in the next section, is reminiscent of a great number of natural processes involving abrupt transitions.

A very convenient representation of the states that can be reached by a system beyond a bifurcation is provided by the *phase space*, the space spanned by the full set of variables $X_1, \ldots, X_n$ participating in the dynamics (Nicolis and Prigogine, 1989; Guckenheimer and Holmes, 1983). The nature of the phase space 'portrait' depends on whether the system is conservative or dissipative. Experiments show that the great majority of systems encountered in nature are dissipative, a property that shows up through an irreversible evolution to a preferred set in phase space, which we call an *attractor*. Attractors enjoy the important property of asymptotic stability, that is to say, the ability to damp perturbations. This in turn ensures a certain degree of reproducibility of the behaviour.

Whatever its detailed nature, an attractor has a measure in phase space which is equal to zero. In other words, in a phase space of n dimensions its dimensionality d will satisfy the strict inequality $d < n$. Actually, in a great number of cases d is much less than n, indicating that a drastic reduction of the description of the system is possible. The simplest attractors

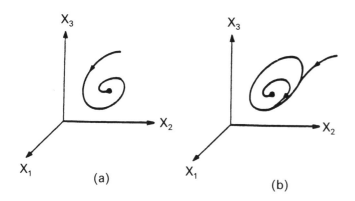

Fig. 2 Evolution towards a point attractor (a) or a limit cycle attractor (b) in a dissipative dynamical system.

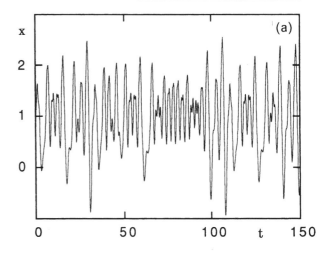

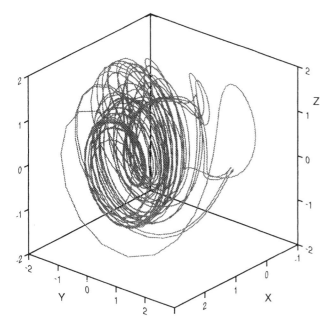

Fig. 3 Chaotic attractor obtained from numerical integration of the system of eq. (2). Parameter values: $a = 0.25$, $b = 4$, $F = 8$, $G = 1.25$.

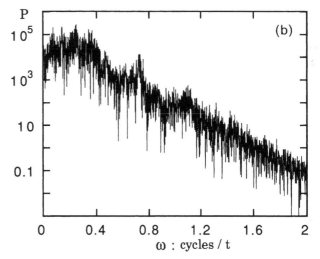

Fig. 4 (a) Time evolution of variable x (eq. (2)) in the regime of deterministic chaos, (b) power spectrum associated to this variable, displaying the broad band character usually attributed to random noise.

$$\frac{dz}{dt} = bxy + xz - z$$

Here x denotes the strength of the westerly wind as affected by large-scale waves whose phases are represented by y and z, whereas the constant terms respresent thermal forcings. All variables are scaled and the time unit is equal to five days, representing the typical damping time of the waves.

Figs. 4(a) and (b) illustrate two ubiquitous features shared by all chaotic attractors: first, contrary to what happens in a periodic, limit cycle attractor (Fig. 2(b)) the trajectory never closes back to itself (Fig. 4(a)); and second, in addition to certain well-defined characteristic frequencies a chaotic attractor generates a broad band spectrum reminiscent of random noise (Fig. 4(b)). Chaotic attractors thus show that irregular, aperiodic behaviour sharing some features of a random process may be generated by a dynamical system governed by perfectly deterministic laws of evolution.

are zero dimensional (point) and one dimensional (limit cycle) manifolds, as depicted in Fig. 2.

However, one now realizes that in many instances bifurcation cascades lead the system to the regime of *deterministic chaos* alluded to already in the Introduction (Baker and Gollub, 1990; Bergé et al., 1986). Fig. 3 depicts a typical attractor describing chaotic dynamics, obtained by numerical integration of a low order, three-variable atmospheric model (Lorenz, 1984):

$$\frac{dx}{dt} = -y^2 - z^2 - ax + aF$$

$$\frac{dy}{dt} = xy - bxz - y + G \qquad (2)$$

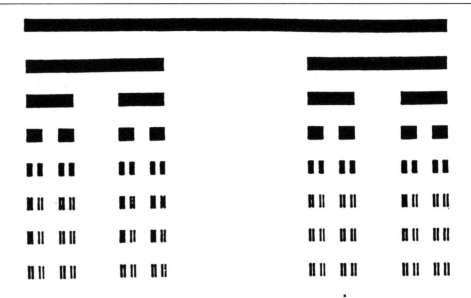

Fig. 5 Successive steps leading to the Cantor set.

Let us have a closer look at the structure of the attractor, Fig.3. We observe two opposing trends: on the one hand an *instability of motion* tending to move the phase space trajectory away from a 'reference state'; on the other hand the bending of the outgoing trajectories followed by their *reinjection* back to the vicinity of this state.

The unstable motion on the attractor (as opposed to the stability in the directions transverse to the attractor) is reflected by the sensitivity of the trajectories on the attractor to minute changes in the initial conditions, as a result of which two initially nearby states diverge, on the average, in an exponential fashion. The characteristic rate of divergence σ_L is referred to as the (largest) positive Lyapounov exponent of the system. For the observer, this property will again be interpreted as a limited predictability. For the model of eq. (2) this entails that beyond σ_L^{-1} time units the system's evolution can no longer be predicted with certainty (Benzi and Carnevale, 1989).

What makes the geometrical object depicted in Fig. 3 and generated by a one dimensional curve, the phase space trajectory, capable of accommodating all this complexity? A detailed study shows that during the different reinjection cycles the attractor undergoes successive foldings. Put differently, a section of the attractor transverse to these folded sheets will look like a line from which an increasing number of segments of decreasing size is removed (Fig. 5).

Present day mathematics provides us with models of this kind of object which are referred to as *fractals*, the particular fractal depicted in Fig. 5 being known as Cantor set (Mandelbrot, 1977). A key mathematical concept characterizing fractal objects is the *correlation dimension v*, expressing essentially the way the number of points N_ε on the attractor at a distance r from a given point varies on the average with r, as

r goes to zero (Grassberger and Procaccia, 1983; Bergé et al., 1986; Baker and Gollub, 1990):

$$N_\varepsilon \approx r^v \qquad (3)$$

Fractals distinguish themselves from traditional objects described by Euclidean geometry by the fact that v is strictly larger than the Euclidean dimension. For instance, in the limit of an infinite number of foldings a Cantor set (Fig. 5) is constituted of an infinity of disconnected points and has thus a Euclidean dimension equal to zero. Yet its fractal dimension turns out to be between 0 and 1. This entails that in the full phase space the attractor of Fig. 3 is a fractal object of dimension between 2 and 3.

A number of alternative definitions of the dimension of fractal objects have been proposed, in order to be able to distinguish between uniform attractors, in which all regions are visited with practically the same probability, and highly nonuniform ones. In general all these dimensions and the Lyapounov exponents σ_L cannot be computed analytically. Algorithms are currently available, allowing the determination of v and σ_L from the knowledge of a time series pertaining to the evolution of a single variable. We shall sketch this procedure in the next section in connection with paleoclimatic data. We close the present discussion by pointing out that the coexistence of the two antagonistic trends of overall stability and reproducibility of the attractor on one side, and of instability of motion of the attractor itself on the other side, make deterministic chaos the natural model for understanding objects whose complexity stems from the coexistence of randomness and order. As pointed out in the Introduction such objects should be important in geosciences. The situation is very different in the presence of

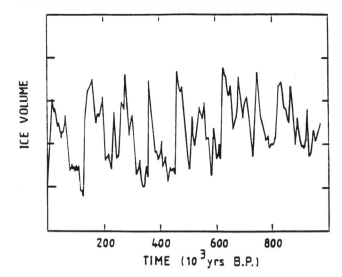

Fig. 6 Global ice volume on earth inferred from oxygen isotope record of core V28238 (after Shackleton and Opdyke, 1973).

random noise, since in this case there is no mechanism capable of keeping the system confined in a privileged part of its state space.

3. CLIMATIC VARIABILITY AND DYNAMICAL SYSTEMS

The quantitative study of a natural system is based on the measurement of a set of relevant variables during a sufficiently long period of time. An important example of such a time series is given by Fig. 6 depicting the oxygen isotope record of a deep sea core during the past 10^6 years (Shackleton and Opdyke, 1973). Such records are thought to provide a reliable estimate of the global ice volume present on the planet earth against time. We observe a number of excursions associated to the glaciation periods, the most dramatic climatic episodes of the Quaternary era. Both the position and strength of these excursions are rather irregular, although on average a time scale of 10^5 years is clearly emerging. In this section we comment on the origin of this variability in the light of the ideas developed above.

The traditional approach in modelling such complex dynamics would consider a linear equation for the relevant variable (here the global temperature or the ice volume) subjected to a random forcing. This type of approach is inadequate for the data depicted in Fig. 6. Indeed, the strength of the noise needed to reproduce the large-scale glacial-interglacial excursions would be exceedingly large.

Another approach consists of the introduction of a periodic component in the above model taking into account the variation of the eccentricity of the earth's orbit, whose amplitude and periodicity are of the order of 10^{-3} and

100 000 years respectively (Berger, 1981). It can easily be verified, however, that the response of such a system is of the order of the amplitude of the forcing itself that is to say, much smaller than the observed response. It seems therefore that simple linear response type models cannot provide us with a plausible mechanism of the Quaternary glaciations. In the language of the previous section this means that the situation in which only one point attractor is available is not representative of the glaciation cycles. We are, therefore, forced to look for an alternative scenario.

The analysis of the preceeding section suggests an alternative origin of the behaviour depicted in Fig. 6, namely that the dynamical system possesses two coexisting attractors and performs intermittent jumps between them (Nicolis and Nicolis, 1981). This implies, automatically, that nonlinear effects are essential in the sense that the dynamics can no longer be expanded around a well-defined 'reference' state and subsequently be linearized to a good approximation.

A simple model corresponding to such a scenario is the globally averaged (also referred to as zero dimensional) energy balance model of the planet earth:

$$\frac{dT}{dt} = v(T,\lambda)$$

$$= \frac{1}{C}\left[(\textit{income solar energy}) - (\textit{outgoing infrared energy})\right]$$

$$= \frac{1}{C}\left[Q((1-\alpha(T)) - \varepsilon_B \sigma T^4\right] \qquad (4)$$

where T is the space averaged surface temperature, C the heat capacity, Q the solar constant, α the albedo, σ the Stefan-Boltzmann constant and ε_B the emissivity factor representing the deviation from black body radiation. The surface-albedo feedback can be readily incorporated in this picture by modelling the albedo as a stepwise linear function. The resulting energy balance is represented in Fig. 7.

For plausible parameter values it can give rise to two stable steady states, T_a and T_b, representing respectively a glacial and a more favourable climate, separated by an intermediate unstable one, T_o. If (as suggested by the record) the difference $T_a - T_b$ is small, the system could be further assumed to operate near a bifurcation point.

Climatic change necessitates a transition between states a and b. Now, in the model elaborated so far no mechanism allowing for such a transition is present except for the trivial one, whereby the system initially at T_b (say) is perturbed and brought near T_a. Such massive perturbations are hard to imagine. We therefore now enlarge our description by incorporating the effect of random fluctuations $F(t)$ which are always present in a complex physical system as a result of the imbalances that inevitably exist between transport or radia-

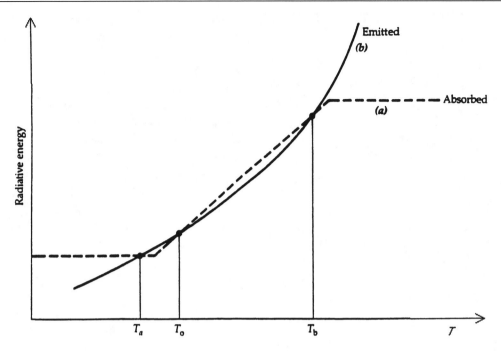

Fig. 7 Incoming and outgoing radiative energy curves as functions of T (global average temperature). Their intersections T_a, T_b and T_o are the steady states predicted by eq. (4).

tive mechanisms. We model these fluctuations as a Gaussian white noise:

$$\langle F(t) \rangle = 0$$
$$\langle F(t)F(t') \rangle = q^2 \delta(t - t') \qquad (5)$$

where q^2 is the variance of fluctuations, and we write the augmented energy balance equation as

$$\frac{dT}{dt} = v(T,\lambda) + g(x)F(t) \qquad (6)$$

where $g(x)$ represents the coupling of the internal dynamics to $F(t)$.

This *stochastic differential equation* (also referred to in the literature as the *Langevin equation*) can be studied by the methods of the theory of stochastic processes. The principal features of the evolution predicted may be summarized as follows. Suppose that the system is started on one of its stable attractors. If the strength of the stochastic forcing $F(t)$ is small, during a long period of time the system will perform a small-scale jittery motion around a level corresponding to this attractor. But sooner or later there is bound to be a fluctuation capable of overcoming the 'barrier' separating this state from the other ones available, in which case over a very short time interval the system will find itself on another attractor. Subsequently it will again undergo a small-scale motion around this new level until a new fluctuation drives it back to the previous state (or to a third one, if available), and so forth. This intermittent evolution looks very much like the

record of Fig. 6. More generally, it provides us with an archetype for understanding a large class of abrupt transition phenomena beyond our specific example like, for instance, recurrent precipitation regimes in some areas, river or lake levels, etc. (Demarée and Nicolis, 1990).

A more quantitative study shows that the transition between attractors occurs on a time scale given by (Gardiner, 1983; Nicolis and Nicolis, 1981)

$$\tau_{a,b} \approx \exp\left\{\frac{2}{q^2}\,\Delta U_{a,b}\right\} \qquad (7)$$

The quantity ΔU_a or ΔU_b is known as the *potential barrier*. It is defined by

$$\Delta U_{a,b} = U_o - U_{a,b} \qquad (8a)$$

where U_o, $U_{a,b}$ is the integral of the right hand side of eq. (4):

$$U(x,\lambda) = -\int^{T} dx\, v(x,\lambda) \qquad (8b)$$

evaluated on states T_o, T_a or T_b. By analogy to mechanics U can be referred to as the *kinetic potential*, since its derivative $\partial U/\partial x$ represents the 'force' v responsible for the evolution. If as expected the variance of the fluctuations q^2 is small and the barrier ΔU finite, τ_a or τ_b will be much larger than the local relaxation time and could well be in the range of glaciation time scales. In contrast, the local evolution in the vicinity of each attractor is given by the inverse of the first derivative of

v, evaluated on T_a or T_b. For an energy balance model it should be, typically, of the order of one year.

Despite the above developed arguments, eq. (6) cannot yet be considered as a satisfactory model of Quaternary glaciations: the transitions between T_a and T_b occur at randomly distributed times, whereas the climatic record suggests that Quaternary glaciations have a cyclic character bearing some correlation with the mean periodicities of the earth's orbital variations. Let us therefore study the response of our multi-stable model to both stochastic and periodic perturbations. Taking the simplest case of a sinusoidal orbital forcing one is led to

$$\frac{dT}{dt} = v(T,\lambda) + g(x)[\varepsilon\sin\omega t + F(t)] \qquad (9)$$

where ε and ω are respectively the amplitude and frequency of the forcing. As pointed out earlier ε is very small, of the order of a fraction of a per cent.

The most striking result pertaining to eq. (9) is, undoubtedly, the possibility of *stochastic resonance* (Nicolis, 1982, 1992; Benzi et al, 1982): when $\Delta U_a \approx \Delta U_b$ and the forcing period $2\pi/\omega$ is of the order of the characteristic passage times $\tau_{a,b}$ the response of the system is dramatically amplified. Specifically, the probability of crossing the barrier is substantially increased and the passage occurs with a mean periodicity equal to the periodicity of the forcing. This allows us to understand, at least qualitatively and despite its weakness, how a periodic forcing may leave a lasting signature in the climatic record.

As we saw in the last section, chaotic dynamics provides a second universal mechanism of evolution in the form of intermittent jumps between different levels. Let us explore this scenario for the paleoclimatic record of Fig. 6. The questions we raise are (a) whether, on the sole basis of the time series data one can identify some intrinsic properties of the dynamics of climate, independent of any modelling, and (b), whether the intermittency and aperiodicity displayed by Fig. 6 are the consequence of deterministic dynamics or of a random course of events impossible to control.

The first step to be carried out in answering these questions is to embed the dynamics of the system under study in phase space. This means, in particular, that one should go beyond the one dimensional view afforded by a time series of a single variable, $X_o(t)$. It can be shown that a phase space satisfying all requirements of dynamical systems theory is the phase space generated by $X_o(t)$ and its successive lags $X_o(t + \tau),\ldots,X_o(t+(n-1)\tau)$. It suffices to choose τ in such a way that these different functions are linearly independent.

Next, in each of the above defined phase spaces (that is to say for each choice of value of n), one tries to identify the nature of the set of data points, viewed as a geometrical object in an n-dimensional space. Again, dynamical systems theory provides algorithms for accomplishing this. One particular quantity which can be identified in this way is the correlation dimension $v(n)$ of our data set (cf. eq. (3)). Once $v(n)$ is determined one can also obtain information on dynamical properties like, for instance, the largest positive Lyapounov exponent $\sigma_L(n)$, if any.

Finally, v and σ_L are plotted against v for increasing values of the embedding dimensionality. If these dependencies are saturated beyond some relatively small n, the system represented by the time series should be a deterministic dynamical system possessing an attractor. The saturation values of v and σ_L will be the dimensionality and the largest Lyapounov exponent of the attractor. The value of n beyond which saturation is observed provides the minimum number of variables necessary to model the behaviour represented by the attractor. If, on the other hand, there is no saturation trend, the conclusion will be either that the system described by the time series evolves in a random way or that the dimensionality is too high to be revealed by the data available.

This procedure, applied to the climatic data of Fig. 6, gives a fractal dimensionality of about $v \approx 3.1$, a positive Lyapounov exponent corresponding to a characteristic predictability time of about 30 000 years and a minimum value of $n = 4$ (Nicolis and Nicolis, 1984, 1986). We are thus allowed to conclude, on the basis of the data, that long term climatic change is to be viewed as a deterministic, unstable dynamics possessing a chaotic attractor. This provides us with a natural way to understand the well-known variability of the climatic system on this long scale. It also allows us to calculate, quantitatively, the limits beyond which predictions about the future become meaningless.

We close with the remark that, far from being contradictory, the last two scenarios of long term climatic change developed in this section present interesting complementarities. Specifically, the picture of a bi-stable system performing fluctuation-driven intermittent jumps between two states may be viewed as a shorthand description of a chaotic attractor which, as stressed previously, shares in many respects the features of a random process. In other words the deterministic part and the random force $v(T, \lambda)$ and $F(t)$ in eq. (6) may represent, respectively, the large-scale structure of the underlying chaotic attractor and the 'effective noise' that is intrinsically generated by the dynamics. In this perspective a more satisfactory approach would be to model $F(t)$ as a highly correlated process rather than as a white noise process. The theory developed in this section can be amended to take this refinement into account, without any substantial change in the overall philosophy underlying the dynamical systems approach to global environmental change.

4. CONCLUDING REMARKS ───────────

We have seen that dynamical systems theory provides us with interesting insights into the variability of our natural environment. We have pointed out that the unstable character of the dynamics of the principal atmospheric and climatic variables entails the existence of an intrinsically generated, irreducible complexity and unpredictability which can in no way be attributed to incomplete knowledge or to the large number of variables and parameters involved.

It is our hope that the ideas set forth in the present paper will contribute to the awareness of the geoscientists about the usefulness of nonlinear dynamical systems, both as a source of inspiration for new ways of looking at sometimes long standing problems and as a quantitative tool of primary importance in the art of modelling and forecasting.

ACKNOWLEDGEMENTS ───────────

This work is supported in part by the Commission of the European Communities and the Federal Office for Scientific, Technical and Cultural Affairs.

REFERENCES ─────────────────

BAKER, G. and GOLLUB J. (1990) Chaotic dynamics. Cambridge Univ. Press, Cambridge, 182 pp.

BENZI, R. and CARNEVALE, F.C. (1989) A possible measure of local predictability. J. Atmos. Sci., 46, 3595–3598.

BENZI, R., PARISI, G., SUTERA, A. and VULPIANI, A. (1982) Stochastic resonance in climatic change. Tellus, 34, 10–16.

BERGÉ P., POMEAU, Y. and VIDAL, C. (1986) Order within chaos. Wiley, New York, 179 pp.

BERGER, A. (Ed.) (1981) Climatic variations and variability: Facts and theories. Reidel, Dordrecht, 795 pp.

DEMARÉE, G.R. and NICOLIS, C. (1990) Onset of Sahelian drought viewed as a fluctuation induced transition. Quart. J. Roy. Met. Soc., 116, 221–238.

GARDINER, C. (1983) Handbook of stochastic methods. Springer-Verlag, Berlin, 442 pp.

GRASSBERGER, P. and PROCACCIA, I. (1983) Characterization of strange attractors. Phys. Rev. Lett., 50, 346–349.

GUCKENHEIMER, J. and HOLMES, P. (1983) Nonlinear oscillations, dynamical systems and bifurcations of vector fields. Springer-Verlag, Berlin, 459 pp.

LORENZ, E.N. (1984) Irregularity: A fundamental property of the atmosphere. Tellus, 36A, 98–110.

MANDELBROT, B. (1977) Fractals: Form, chance dimension. Freeman, San Francisco, 365 pp.

NICOLIS, C. (1982) Stochastic aspects of climatic transitions – Response to a periodic forcing. Tellus, 34, 1–9.

NICOLIS, C. (1993) Long term climatic transitions and stochastic resonance. J. Stat. Phys., 70, 3–14.

NICOLIS, C. and NICOLIS, G., (1981) Stochastic aspects of climatic transitions – Additive fluctuations. Tellus, 33, 225–234.

NICOLIS, C. and NICOLIS, G. (1984) Is there a climatic attractor? Nature, 311, 529–532.

NICOLIS, C. and NICOLIS, G. (1986) Reconstruction of the dynamics of the climatic system from time series data. Proc. Nat. Acad. Sci. US, 83, 536–540.

NICOLIS, C. and NICOLIS, G. (Eds) (1987) Irreversible phenomena and dynamical systems analysis in geosciences. Reidel, Dordrecht, 578 pp.

NICOLIS, G. and PRIGOGINE, I. (1989) Exploring complexity. Freeman, New York, 313 pp.

SHACKLETON N.J. and OPDYKE, N.D. (1973) Oxygen isotope and paleomagnetic stratigraphy of equatorial Pacific core V28–238; oxygen isotope temperature and ice volumes on a 10^5 and 10^6 year scale. Quatern. Res., 3, 39–55.

From scalar cascades to lie cascades: joint multifractal analysis of rain and cloud processes

D. SCHERTZER[1] and S. LOVEJOY[2]

[1] *Laboratoire de Métérologie Dynamique (CNRS), case 99,*

Université P&M Curie, 4 Pl. Jussieu,

Paris 75252 Cedex 05,

France

[2] *Department of Physics, McGill University,*

3600 University St.,

Montreal, Quebec,

Canada, H3A 2T8

ABSTRACT   There are two primary approaches to modeling rainfall; stochastic modeling and deterministic integration of nonlinear partial differential equations which model the atmospheric dynamics. The statistical advantages of the former could be combined with the physical advantages of the latter by exploiting cascade models based on scale invariant symmetries respected by the equations. Carried to its logical conclusion, this approach involves considering the atmosphere as a space-time multifractal process admitting either a vector, tensor or even only a nonlinear representation. The process is then defined by two groups which respectively specify the rule required to change from one scale to another and the corresponding transforms of fields. Both groups are characterized by their generators, hence by their Lie algebra. We show how to extend existing cascades beyond scalar processes, showing preliminary numerical simulations and data analyses, as well as indicating how to characterize and classify the scale invariant interactions of fields.

1. INTRODUCTION

1.1 The limitations of standard deterministic dynamical and of phenomenological stochastic modeling of rain

Geophysical fields show abundant evidence of nonlinear variability resulting from strong nonlinear interactions between different scales, different structures, and different fields. This variability is quite extreme and is associated with catastrophic events such as earthquakes, tornadoes, flash floods, extreme temperatures, volcanic eruptions. Another fundamental characteristic of this variability is the very large range of scales involved, which often extends from 10,000 km to 1 mm in space, and from geological scales to milliseconds in time. The scale ratio associated with this variability is at least 10^9, and for geophysical flows the corresponding Reynolds number is typically of the order of 10^{12} – so large that

without any doubt the dynamics are all turbulent. Recently, a systematic study (Lovejoy et al., 1993) of scaling of cloud radiances at visible and infrared wave lengths (see Fig. 1) has revealed that as suggested by the unified scaling model of atmospheric dynamics (Schertzer & Lovejoy, 1983, 1985) – the scaling holds over at least the range ≈ 4000 km to ≈ 300 m (see also recent dynamics studies (Chigirinskaya et al., 1994, Lazarev et al., 1994)).

Up until recently, there have been two primary approaches to rainfall modeling: phenomenological stochastic modeling favoured by hydrologists, and deterministic dynamical modeling favoured by meteorologists. The former was largely based on ad hoc methods designed to mimic a phenomenology associated notably with a group at the University of Washington (e.g. Austin & Houze, 1972) that is predicated on the assumption that rain processes are qualitatively different a factor two or so in scale. The scientific

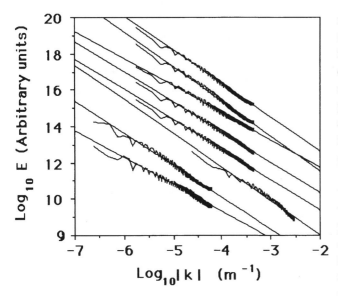

Fig. 1 Average power spectrum for the satellites' images grouped according to the satellite and the frequency range of the images (from bottom to top): LANDSAT (visible) $\beta = 1.7$, METEOSAT (visible) $\beta = 1.4$, METEOSAT (infrared) $\beta = 1.7$, NOAA-9 (channel 1 to 5) $\beta = 1.67$, 1.67, 1.49, 1.91, 1.85 (from Lovejoy et al., 1993).

outcome of relying on this phenomenology has been a series of very complex cluster processes where hierarchies of time and space scales are each assigned 'plausible' variations in rain rate and statistical fluctuations. Perhaps the best known model of this type is the Waymire-Gupta-Rodriguez-Iturbe 1984 (WGR) model which involves a dozen or so empirical parameters, and is at best successful only within the narrow range of time and space scales for which it was calibrated. The variability of the smaller scales not explicitly incorporated into the model yields a behaviour which is unrealistically smooth. The variability of the larger scales yields a behaviour which has unrealistically small variation from storm to storm. A further criticism of this approach that we will outline here is based on the fact that the rain process is coupled to other atmospheric fields in a highly nonlinear way; it cannot be fully treated in isolation. Consequently, rain should really be regarded as a component field of a space-time vector process where each component of the vector represents a different interacting field. Phenomenological stochastic models of the WGR type can be neither trivially nor naturally extended to include these other interacting fields.

In contrast, the deterministic models were developed following the usual methods of geophysical fluid dynamics. They are predicated on the integration of nonlinear partial differential equations which attempt to represent the complex nonlinear dynamics including a hopefully appropriate parametrization of the 'physics'. Because of the limited number of degrees of freedom which can be explicitly

modeled, this approach makes drastic scale truncations (studying one scale independently of the others), transforming partial differential equations into ordinary differential equations, arbitrarily hypothesizing the homogeneity of subgrid scale fields, and performing ad hoc and unjustified parametrizations. In summary, both of these traditional approaches are therefore fundamentally limited by their inability to deal adequately with variability spanning many orders of magnitude in scale.

Even if we ignore these (over-) 'simplifying' assumptions, the consequences of such choices (which have increasingly weak links with the real world) are ultimately complex and yield unwieldy numerical codes. The relevance of such codes, obtained after either this long series of butcherings of the initial equations or after a long series of ad hoc attempts to mimic the phenomenology, remains highly questionable. For example, there is an increasing tendency to test deterministic models by making intercomparisons with other models! In contrast, the phenomenological stochastic approach does make closer contact with the data, but is virtually useless outside of the narrow range within which it is calibrated. Moreover, it is not even able to deal with extreme events within the calibration range. Both of these approaches suffer from strong limitations due to their inability to come to grips with the fundamental problem of nonlinear variability. This problem must be overcome if we want eventually to understand the very noisy intermittency of the signals of hydrology and other geophysical systems.

1.2 Cascades and symmetries

An alternative approach to nonlinear variability – first clearly elaborated by Schertzer & Lovejoy (1987) – is based on a fundamental symmetry property of the nonlinear (e.g. Navier-Stokes) equations: scale invariance. Indeed, the simplest way of understanding how geophysical variability occurs over a very large range of scales is to suppose that the same type of elementary process acts at each relevant scale (from the large scale down to the small viscous scale). At first, this began as a fractal approach (even before the word was coined, e.g. Richardson's (1922) celebrated poem on self-similar cascades), then (after 1983) it evolved into a multifractal approach.

These scale invariant multifractal models are superficially quite simple phenomenological 'toy models' (a bit like cellular automata). They give rise to cascades, avalanches, and other exotic phenomena (exotic compared to conventional smooth mathematical descriptions of the real world), but nevertheless have highly nontrivial consequences! For example, as we will see later, simple cascade models already give rise to a fundamental difference between observables and truncated processes, and such a difference is a general property of the wide class of 'hard' multifractal processes (which distinguish between 'dressed' and 'bare' properties respectively). These models produce hierarchies of self-organized random structures, which is also a very general property of (singular) multifractal measures of singularities and of Self-Organized Criticality (Bak et al., 1987). In general, these simple models give us precious hints as to how to cast order in disorder.

Until now, a basic limitation of these cascade processes is that they have been limited to positive scalar fields (such as the energy flux from large to small scales). They are thus even capable of dealing with inverse cascades (negative fluxes; small to large scale transfers), not to mention the more fundamental problem of vector (e.g. wind) or tensor processes (e.g. stress or strain tensors) necessary to deal fully with the nonlinear dynamics. Below, we give a brief review of this scalar cascade theory and then go on to show how it can be generalized to vector, tensor or 'Lie cascades'.

2. THE SCALAR MULTIFRACTAL FRAMEWORK

2.1 Fractals and multifractals: Fractal geometry

Fractal geometry provides the simplest nontrivial example of scale invariance, and is useful for characterizing fractal sets. It can also be useful in producing linear models of rain produced by additive random processes (which involve unique fractal dimensions) such as the (monofractal) simple scaling model of rain tested in Lovejoy (1981) (see also Lovejoy & Mandelbrot, 1985; Lovejoy & Schertzer, 1985). Unfortunately in geophysics we are much more interested in fields and are rarely interested in geometrical sets. However, fractal dimensions can still be useful in counting the occurrences of a given phenomenon over a wide range of scales – as long as we can properly pose this question. If this is the case and the phenomenon is scaling, then the number of occurrences ($N_A(l)$) of an event at resolution scale l (in space and/or time) follows a power law[1]:

$$N_a(l) \approx \left(\frac{l}{L}\right)^{-D_F} \tag{1}$$

D_F being the (unique) fractal dimension, generally not an integer, and L the (fixed) largest scale. For instance, Fig. 2 shows the records of rain events during the last 45 years in Dedougou (Hubert & Carbonel, 1990). These authors show that the occurrence of rainy days during a certain time scale T is fractal, having a dimension $D_F \approx 0.8$, which accounts for the fact that the rain events on the time axis form a Cantor-

like set. Amusingly, the wet season is often considered to last 7 months per year, and $0.8 \approx \text{Log}7/\text{Log}12$ (recall that the standard Cantor set is obtained by iteratively removing the (closed) middle section of the unit interval and has a dimension of $\text{Log}2/\text{Log}3$).

Numerous similar (mono-) fractal results can be obtained on different fields. However, fields having different levels of intensity do not reduce to the oversimplified binary question of occurrence or nonoccurrence. For instance, in the case of rain we have to address the fundamental question: what is the rain rate at different scales? What is a negligible rain rate? Generalizations of fractal/scale invariance ideas well beyond geometry were desperately needed and appeared in 1983 when the dogma of a unique dimension was finally abandoned (Henstchel & Procaccia, 1983; Grassberger, 1983; Schertzer & Lovejoy, 1983).

However, it is already important to note that the notion of codimension (c) (usually defined by $c = D - D_F$, where D is the dimension of the embedding space) can be considered to be at least as fundamental as the notion of fractal dimension D_F. Indeed, c can be directly defined as measuring the fraction of the space occupied by the fractal set A of dimension D_F. This can be seen by considering that a ball B_l of size l has the following probability of intersecting A:

$$P(B_l \cap A) \approx \frac{N_A(l)}{N(l)} = l^c \tag{2}$$

where $N(l) \approx l^{-D}$ is the number of balls size l necessary to cover a D-dimensional space.

In fact, for multifractal fields, codimensions will be more fundamental and useful than dimensions, since they give intrinsic characterizations of the multifractal process. We will therefore use a codimension formalism (Schertzer & Lovejoy, 1987, 1992) rather than the more popular dimension formalism developed for strange attractors (e.g. Halsey et al., 1986). One may note that recently the need of a codimension formalism has been implicitly acknowledged in Mandelbrot (1991).

2.2 The extension to scalar multifractals

One obtains much more information by looking not at the occurrence of rain, but at the rain rate: a 1 mm daily rain rate is negligible compared to a 150 mm daily rain rate! For instance, Fig. 3 displays the rain rate at Nîmes (France) during a few years, and averaged over varying scales T (from a day to a year). This figure illustrates the great intermittency of rain rates: most of the time it is negligible, while sometimes it reaches 200 mm (228 mm in few hours, for the famous October 1988 catastrophe!) – in comparison the daily average is ≈ 2.1 mm. The variability is so significant in this time series that Ladoy et al. (1991) found some evidence of the

---

[1] Here and below the sign $\approx$ means equality within slowly varying and constant factors.

Fig. 2 Picture of 45 years of daily rain rates in Dedougou (Burkina Fasso). Each line corresponds to one year of observation, and each black dot to a rainy day. (Hubert & Carbonnel 1990). The rain events form a Cantor-like set of dimension $D_F \approx 0.8$ (the standard Cantor set is of dimension Log(2)/Log(3)).

Fig. 3 Rain rates in Nîmes (France) during the years 1978–1988, also averaged over 1 day, 4 days, 16 days, 64 days, 256 days, 1024 days and 4096 days respectively (after Ladoy et al., 1993). It illustrates the great intermittency of rain rates: some rare but extreme events of short periods (singularities) gave overwhelming contributions (e.g. the 228 mm which occurred in a few hours) to the October 1988 Nîmes catastrophe, which are hardly smoothed over longer periods.

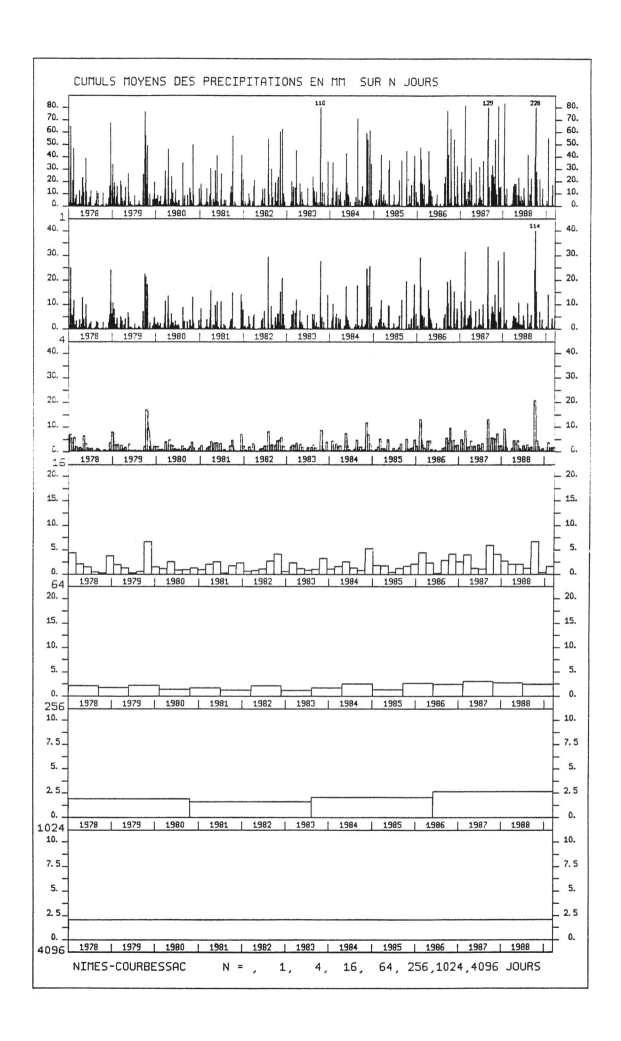

CUMULS MOYENS DES PRECIPITATIONS EN MM SUR N JOURS

NIMES-COURBESSAC   N = ,   1,   4,   16,   64, 256,1024,4096 JOURS

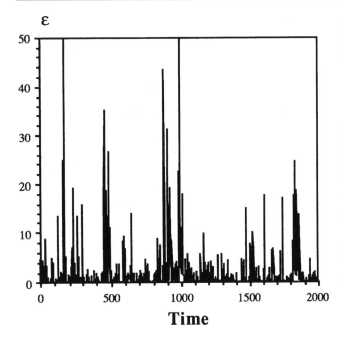

Fig. 4 A representation of the rate of energy transfer ε, computed from experimental hot wire wind measurements (after Schmitt et al., 1992b). The intermittent nature of atmospheric turbulence is obvious.

divergence of moments (a subject we will discuss more below). Qualitatively this variability seems strikingly analogous to that of the energy flux cascade in turbulence (as displayed in Fig. 4), an analogy which turns out to be quite profound.

MULTIPLICATIVE CASCADE PROCESSES

It has become clear that the process transferring energy from larger to smaller scales in turbulence is a multiplicative process (Kolmogorov, 1962; Obukhov, 1962; Yaglom, 1966; Novikov & Stewart, 1964; Mandelbrot, 1974): a random factor determines the fraction of the rate of energy transferred from one large eddy to one of its subeddies (as shown on Fig. 5 for a two-dimensional cut). Larger structures are thus multiplicatively modulated by smaller ones. If we iterate this construction infinitely (the scale ratio $\lambda = L/l$ where L is the larger scale and l the smallest resolved scale which corresponds to the spatial resolution of our field), as $\lambda \to \infty$, we observe singularities: at some points the field diverges (a singularity), whereas over most of the space it goes to zero (a regularity; for convenience we collectively call both behaviours 'singularities').

The simplest multifractal model of this genre is the α-model (Schertzer & Lovejoy, 1983; Levich & Tzvetkov, 1985; Bialas & Peschanski, 1986; Levich & Shtilman, 1991) obtained with a random two-state multiplicative factor: the only restriction is conservation of the ensemble average. If after n iterations of the multiplicative process, ε_n is the value

of the field (for example, the energy flux to smaller scales), we have the relation $\varepsilon_n = \mu\varepsilon\, \varepsilon_{n-1}$, where $\mu\varepsilon$ is a random variable which can have two values with the probabilities:

$$\begin{aligned} \Pr(\mu\varepsilon\lambda^{\gamma^+}) &= \lambda^{-c} \quad \textit{strong} \quad \text{subeddies } (\gamma^+ > 0) \\ \Pr(\mu\varepsilon\lambda^{\gamma^-}) &= 1 - \lambda^{-c} \quad \textit{weak} \quad \text{subeddies } (\gamma^- < 0) \end{aligned} \quad (3)$$

γ^+, γ^-, c are constrained so that the ensemble average $\langle\mu\varepsilon\rangle = 1$ (ensemble average conservation of ε). The (monofractal) β-model (Frisch et al., 1978) is obtained when $\gamma^- = -\infty$, $\gamma^+ = c$: the subeddies are dead or alive, c is the codimension of the (unique) support of turbulence of dimension $D_F = D - c$.

BARE DRESSED QUANTITIES

Figs. 6 and 7 show examples (in 1 and 2 dimensions) of an α-model developed from a large scale to a small homogeneity scale: the 'bare' cascade, which develops singularities when the homogeneity scale goes to zero. But our sensors (e.g. satellites) have not such a small resolution (the homogeneity scale is perhaps of the order of millimeters), and what we observe is an averaged field; these are the 'dressed' quantities in the sense that the observation hides, dresses the activity occurring on scales smaller than that of the observation. On the contrary, in the same sense, a cascade whose development is limited to the scale l is 'bare' on this scale of observation: no smaller scale activity is hidden. Figs. 6 and 7 show that the small scale singularities which appear when we develop the cascade to the homogeneity scale may give overwhelming contributions to the larger scale fluctuations of the dressed quantities. These contributions can be so important that as we will see later, they may imply divergence of higher order statistical moments, corresponding to 'hard' multifractal behaviour.

MULTIFRACTAL FIELDS

In the α-model, pure singularities γ^-, γ^+ (only when $\gamma > 0$ do we obtain singularities, otherwise they are rather regularities, but for convenience all the γ are called 'singularities') give rise to an infinite hierarchy of mixed singularities ($\gamma^- \le \gamma \le \gamma^+$) ($n^{\text{th}}$ step, $\lambda_1 =$ step scale ratio):

$$\gamma = \frac{n^+\gamma^+ + n^-\gamma^-}{n}; \ n^+ + n^- = n; \\ Pr(n^+ = k) = C_n^k \lambda_1^{-ck}(1 - \lambda_1^{-c})^{n-k} \quad (4)$$

The probability density of the resulting field is given by:

$$Pr(\varepsilon_n \ge (\lambda_1^n)^\gamma) \approx (\lambda_1^n)^{-c_n(\gamma)} \quad (5)$$

When $n > > 1$: $c_n(\gamma) \approx c(\gamma)$, a function independent of n, and the probability density of the multifractal turbulent field ε_λ (the field ε at any scale ratio λ) is given by Schertzer & Lovejoy (1987b):

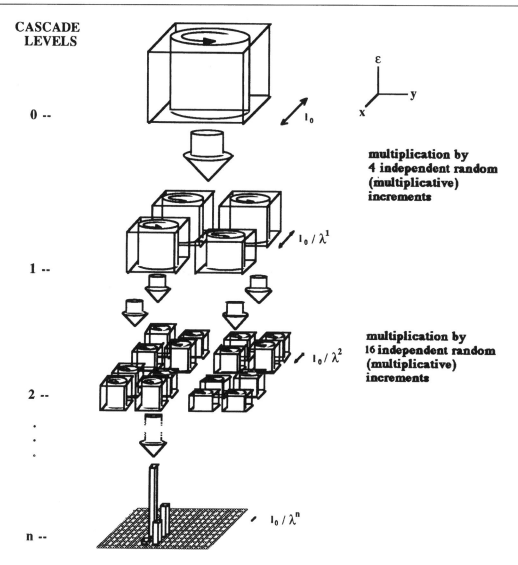

CASCADE LEVELS

multiplication by
4 independent random
(multiplicative)
increments

multiplication by
16 independent random
(multiplicative)
increments

Fig. 5 A schematic diagram showing a two-dimensional cascade process at different levels of its construction to smaller scales. Each eddy is broken up into four subeddies, transferring a part or all its energy flux to the subeddies. In this process the flux of the field at large scale multiplicatively modulates the various fluxes at smaller scales; the mechanism of flux redistribution is repeated at each cascade step (self similarity).

$$Pr(\varepsilon_\lambda \geq \lambda^\gamma) \approx \lambda^{-c(\gamma)} \qquad (6)$$

When $c(\gamma) < d$ (d being the dimension of space), as already discussed (eq. (2)), $c(\gamma)$ is the (geometrical) codimension $c(\gamma) = d - D(\gamma)$ corresponding to the (geometrical) fractal dimension $D(\gamma)$ of the support of singularities whose order is greater than γ.

In the most interesting cases $c(\gamma) \geq d$ is associated with a nonobviously negative (!) dimension $D(\gamma)$ (Mandelbrot, 1991). However the function $c(\gamma)$ remains a (finite) codimension on an (infinite dimensional) probability space (see below). The multiple scaling behaviour of this field ε at scale ratio λ can be also characterized by the corresponding law for the statistical moments (via a Laplace transform):

$$\langle \varepsilon_\lambda^q \rangle \approx \lambda^{K(q)} \qquad (7)$$

The relations between the turbulent notation used here and the strange attractor $f_D(\alpha_D)$ and $\tau_D(q)$ notation (the subscript D explicitly emphasizes the dependence of α, f, τ on the dimension of the observing space D) are: $f_D(\alpha_D) = D - c(\gamma)$ and $\tau_D(q) = K(q) - (q-1)D$ with $\alpha_D = (D - \gamma)$. The codimension notation is necessary when dealing with stochastic processes because γ, c, K are intrinsic contrary to α_D, f_D, τ_D which diverge with $D \to \infty$. It also avoids introducing negative ('latent') dimensions when $c(\gamma) > D$.

Just as $f(\alpha)$ is the Legendre transform (Parisi & Frisch, 1985) of $\tau(q)$, so $c(\gamma)$ is the transform of $K(q)$:

$$K(q) = \max_\gamma (q\gamma - c(\gamma)); \qquad c(\gamma) = \max_q (q\gamma - K(q)) \qquad (8)$$

These relations establish a one-to-one correspondence between orders of singularities and moments ($q = c'(\gamma)$, $\gamma = K'(q)$).

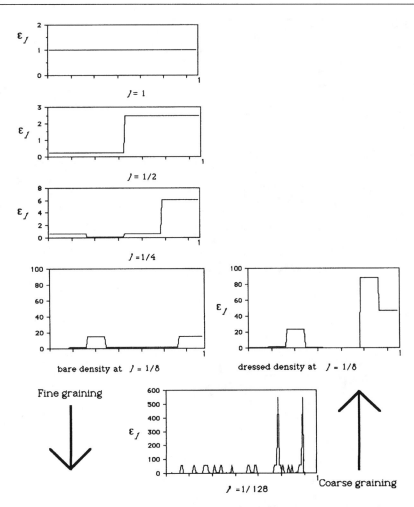

Fig. 6 Illustration (Schertzer and Lovejoy, 1989) of the 'bare' and 'dressed' energy flux densities. The left hand side shows the step by step construction of a 'bare' multifractal cascade (the α-model) starting with an initially uniform unit flux density. The vertical axis represents the density of energy ελ flux to smaller scale with its ensemble average being conserved $\langle \varepsilon_\lambda \rangle = 1$. At each step the horizontal scale is divided by two. The developing spikes are incipient singularities of various orders (characteristic of multifractal processes).

2.3 Some basic properties of multifractal fields

Multifractal fields, contrary to (mono-) fractal geometry, involve an infinite hierarchy of γs corresponding to the infinite hierarchy of $c(\gamma)$. Indeed, according to eq. (5), the hierarchy of codimensions may be obtained by thresholding the field and computing the fractal codimension of values greater than this threshold λ^γ (see Fig. 8). The codimension function $c(\gamma)$ must satisfy only a rather weak constraint (see Fig. 9): not only should it be obviously an increasing function of γ (if $\gamma_1 > \gamma_2$, $\Pr(\varepsilon_\lambda \geq \lambda^{\gamma_1}) \leq \Pr(\varepsilon_\lambda \geq \lambda^{\gamma_2})$, thus $c(\gamma_1) \geq c(\gamma_2)$), but it must also be convex as $K(q)$ (Feller, 1971).

THE SAMPLING DIMENSION D_s

Here we point out the utility of the notion of *sampling dimension D_s*. As we are always compelled to analyze finite samples, it is rather obvious that the highest singularities will rarely be present in a given sample. More precisely speaking, some of the singularities will almost surely not be present in a finite sample. Indeed, when we analyze only one sample/

realization of the field on a dimension D at resolution λ, the largest singularity γ_s we can reach is given by $c(\gamma_s) = D$. More generally, if we are studying N_s samples, we can introduce the sample dimension $D_s \neq 0$ (at resolution λ) defined as $N_s \approx \lambda^{D_s}$ (Schertzer & Lovejoy, 1989; Lavallée, 1991; Lavallée et al., 1991). This largest singularity increases with D_s, since its order is then given by $c(\gamma_s) \approx D + D_s$ (see Fig. 10), it corresponds to a moment order $q_s = c'(\gamma_s)$ beyond which $K(q)$ becomes spuriously linear. The sampling dimension D_s gives us a quantitative way to describe how larger samples enable us to explore more and more of the probability space, eventually attaining the rare singularities responsible for the wild behaviour of experimental fields.

CLASSIFICATION OF MULTIFRACTAL FIELDS

Most of the theoretical and corresponding empirical studies unfortunately presuppose very restrictive calmness and regularity assumptions on multifractal field. Such a limited view

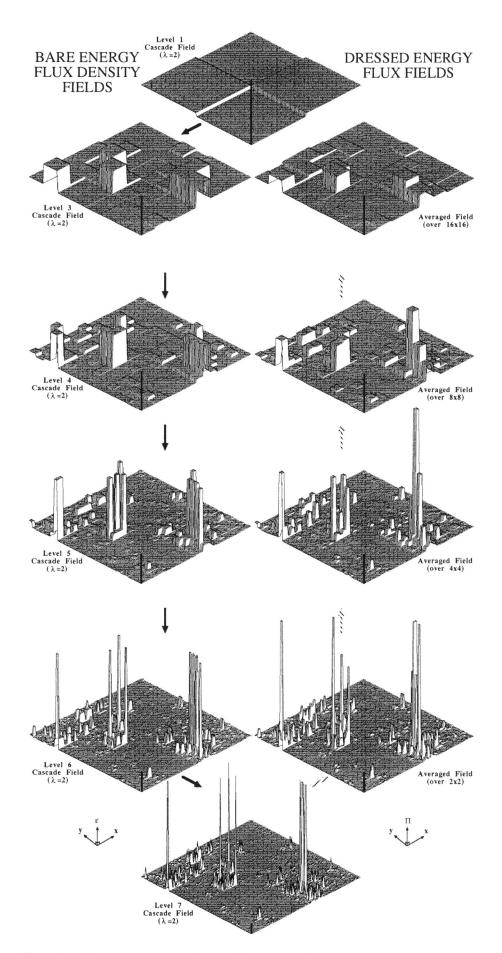

Fig. 7 As in fig. 6, illustration of the 'bare' and 'dressed' energy flux densities, but now for a two-dimensional cut.

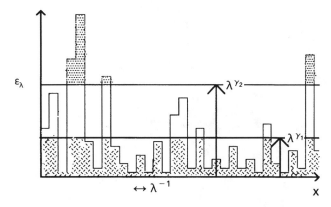

Fig. 8 The importance of the threshold to analyze a multifractal field: if $\gamma_1 < \gamma_2$, $D(\gamma_1) > D(\gamma_2)$, and $c(\gamma_1) < c(\gamma_2)$, the codimension function is an increasing function.

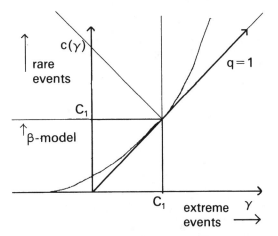

Fig. 9 A schematic $c(\gamma)$ diagram. For conservative process, there is a stationary value (C_1).

of multifractals is quite misleading. It is therefore a matter of some importance to reveal the full diversity of multifractality and classify the different types of possible multifractal field. Indeed a purely geometric approach (without any reference to a stochastic process; Parisi & Frisch, 1985; Paladin & Vulpiani, 1987) presupposes that the singularities are bounded by $\gamma_{max}^{(g)}$, the upper bound of geometrical singularities, with $c(\gamma_{max}^{(g)}) = D$ and $\gamma_{max}^{(g)} < D$. Stochastic processes are generally capable of having singularities of all orders (i.e. $c(\gamma) \geq D$). However, conservation of the flux (e.g. energy flux in turbulence) may introduce a new constraint, which will depend on the type of conservation involved. If we assume microcanonical conservation (i.e. conservation on each realization, see Benzi et al., 1984; Pietronero & Siebesma, 1986; Meneveau & Sreenivasan, 1987[2]; Sreenivasan & Meneveau, 1988), then the singularities are bounded above by $\gamma_{max}^{(m)} = D$. The superscript m corresponding to 'microcanonical', this

---

[2] Their celebrated 'p model' is in fact nothing more than a microcanonical restriction of the α-model discussed earlier.

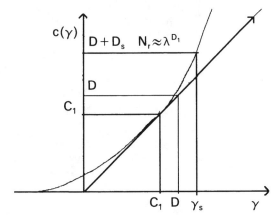

Fig. 10 With a larger and larger number of samples N_s, the maximum reachable singularity γ_s increases ($N_s = \lambda^{Ds}$; $c(\gamma_s) = D + D_s$; Schertzer & Lovejoy, 1989).

bound is reached only for the extreme case in which for each step all the density of the flux is concentrated on a single subeddy (of volume λ^{-D}). We may say that singularities remain 'calm' and 'soft'. As soon as we leave this restricted framework to canonical conservation (i.e. ensemble conservation) we have some 'wild' singularities $\gamma > \gamma_{max}^{(m)}$, which can even be 'hard' in the sense that they are responsible of the divergence of moments (see below).

DIVERGENCE OF MOMENTS, MULTIFRACTAL PHASE TRANSITIONS AND SELF-ORGANIZED CRITICALITY

It is possible to show (Schertzer & Lovejoy, 1987) that the integration of ε_λ^q over a set of dimension D diverges (for $\lambda \to \infty$) when $q > q_D$; q_D given by: $K(q_D) = (q_D - 1)D$; $q_D > 1$. More precisely, consider the flux over a volume element B_λ (scale ratio λ):

$$\Pi(B_\lambda) = \int_{B_\lambda} \varepsilon d^D x \qquad (9)$$

where $\varepsilon = \lim_{\lambda \to \infty} \varepsilon_\lambda$. If we now consider $\varepsilon_{\lambda,D} = \Pi(B_\lambda)/\text{Volume}(B_\lambda)$ as a (dressed) estimate of the (bare) ε_λ over the D-dimensional ball B_λ; the two will have totally different statistical properties:

$$\langle \varepsilon_\lambda^q \rangle < \infty \; all \; q; \qquad \langle \varepsilon_{\lambda,D}^q \rangle = \infty \; q \geq q_D(>1) \qquad (10)$$

This is the fundamental difference between the two quantities 'bare' ε_λ and 'dressed' $\varepsilon_{\lambda,D}$. One may note that the singular statistics (of dressed quantities) has been taken as a basic feature of self organized criticality (Bak et al., 1987). Divergence of moments of a random variable X ($< X^q > = \infty$ for $q > q_D$) corresponds to a 'hyperbolic' (algebraic) fall-off of the probability distribution:

$$Pr(X \geq s) \approx s^{-q_D}(s \gg 1) \Leftrightarrow \langle X^q \rangle = \infty \quad for \; q > q_D \quad (11)$$

The physical significance of divergence of moments is that when $q < q_D$ the dressed moments are macroscopically determined whereas for $q > q_D$ the moments will be microscopically determined depending crucially on the small scale details. It is possible to make a formal[3] analogy between conventional thermodynamics and multifractals; for example, the entropy corresponds to $c(\gamma)$ and the temperature to $1/q$, the Massieu potential (the free energy divided by temperature) to $K(q)$. Therefore, this qualitatively new behaviour for $q > q_D$ (low temperatures) can be considered as discussed in Schertzer & Lovejoy (1992, 1994) and Schertzer et al. (1993); this corresponds to a first order multifractal phase transition, where the thermodynamic potential $K(q)$ has a first order discontinuity at the critical temperature analog q_D^{-1} (Schmitt et al. 1994; Chigirinskaya et al. 1994 for corresponding atmospheric data analysis: $q_D \simeq 7$ for velocity field).

2.4 The three fundamental exponents: H, C_1, α

It is already important to note that three parameters are sufficient to characterize *locally* (around the mean singularity) the infinite hierarchy of fractal codimensions $c(\gamma)$. Furthermore, this characterization turns out to be global under certain general hypotheses of universality we discuss in the next section. The three fundamental exponents are the following:

- H describes the *deviation from conservation* of the flux: $\langle \varepsilon_\lambda \rangle \approx \lambda^{-H}$. $H = 0$ for conservative fields (for instance the energy flux in turbulence, $\langle \varepsilon_\lambda \rangle$ independent of λ) whereas according to the Kolmogorov relation in real space $\Delta v_\lambda \approx \varepsilon_\lambda^{1/3} \lambda^{-1/3}$ (where Δv_λ is the wind shear amplitude $|v(x + \lambda^{-1}) - v(x)|$ at scale ratio λ), the wind shear is a nonconservative field ($H = 1/3$).
- C_1 describes the *mean inhomogeneity*: it is the codimension of the mean singularity: $C_1 = c(C_1 - H)$, in the case of conservative fluxes it is also the order of the mean singularity (and simultaneously the fixed point of $c(\gamma)$).
- α represents the *degree of multifractality* measured by the convexity of $c(C_1)$ around the mean singularity $(C_1 - H)$ measured by the radius of curvature: $R_c(\gamma = C_1 - H) = 2^{3/2} \alpha C_1$ which increases with the range of singularities (starting from zero with the monofractal β-model). As shown below, in the case of universal multifractals, α is also the Levy index of the generator and $0 \leq \alpha \leq 2$.

2.5 Universality by mixing of multifractal processes

The particularities of the discrete models (e.g. α-model) remain as the cascade proceeds to its small scale limit. If we simply iterate the model step by step with a fixed ratio of scale λ_1 for the elementary step, we indefinitely increase the range

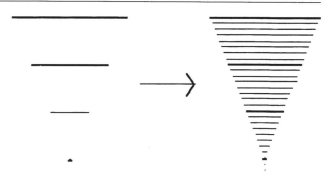

Fig. 11 Keeping the total range of scale fixed and finite, introducing intermediate scales and then seeking the limit of an infinite range of scales leads to universality.

of scales $\Lambda \to \infty$ which already poses a nontrivial mathematical problem (weak limit of random measures, see Kahane, 1985). On the contrary, by *keeping* the total *range of scale fixed and finite, mixing independent processes* of the *same type* (by multiplying them, preserving certain characteristics, e.g. variance of the generator) and *then* seeking the limit $\wedge \to \infty$, *a totally different limiting problem is obtained!* For instance, this may correspond to *densifying* the excited scales by introducing more and more intermediate scales (see Fig. 11), and seeking thus the limit of continuous scales of the cascade model. Alternatively, we may also consider the limit of multiplications of i.i.d. discrete cascades models.

In both cases, multiplying processes corresponds to adding generators: $\varepsilon_\lambda \approx e^{\Gamma_\lambda}$ where ε_λ is the process and Γ_λ is the generator. If we seek *generators* which are *stable* and *attractive* under *addition* (using the results on the second Laplace characteristic function $K(q)$ equivalent to the free energy), we must consider (Schertzer & Lovejoy, 1987, 1989; Schertzer et al., 1988; Fan, 1989) stable extremal Lévy noises with $1/f$ spectra, which are characterized by a *Lévy index* α: $Pr(-\Gamma \geq s) \approx s^{-\alpha}$ $(s \gg 1)$ $\Rightarrow$ any $q \geq \alpha$: $\langle (-\Gamma)^q \rangle = \infty$. Except for the Gaussian exception $\alpha = 2$, α is the order of divergence of moments of the generator. These generators yield the following *universal* expressions for the scaling function of the moments of the field $K(q)$ and of the codimension function $c(\gamma - H)$:

$$c(\gamma - H) = C_1 \left(\frac{\gamma}{C_1 \alpha'} + \frac{1}{\alpha} \right)^{\alpha'} \qquad \alpha \neq 1$$

$$c(\gamma - H) = C_1 \exp \left(\frac{\gamma}{C_1} - 1 \right) \qquad \alpha = 1 \tag{12}$$

$$K(q) + Hq = \frac{C_1}{\alpha - 1} (q^\alpha - q) \qquad \alpha \neq 1$$

$$K(q) + Hq = C_1 q Log(q) \qquad \alpha = 1 \tag{13}$$

where $(1/\alpha + 1/\alpha' = 1$, and for $q = dc/d\gamma > 0)$ and C_1 is related to the coefficient C of the canonical Lévy measure dF by:

---

[3] Formal since we here are considering systems out of equilibrium.

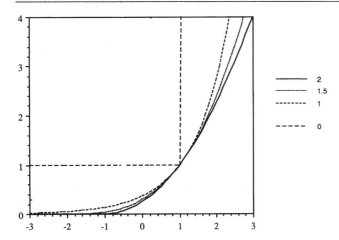

Fig. 12 Universal (bare) singularities codimension $c(\gamma)/C_1$ corresponding to the five classes; here $\alpha = 2, 1.5, 1, 0$.

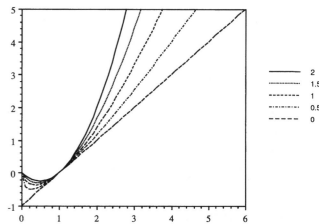

Fig. 13 Universal (bare) singularities codimension $c(\gamma)/C_1$ corresponding to the five classes; here $\alpha = 2, 1.5, 1, 0.5, 0$.

$$C_1 = C\Gamma(3-\alpha)/\alpha; \quad dF = 1_{x>0}C(2-\alpha)x^{-\alpha}dx/x \quad (14)$$

(here Γ is the usual gamma function and should not be confused with the generator. Figs. 12 and 13 show universal $K(q)$ and $c(\gamma)$ curves, α varying from 0 to 2.

The two functions $K(q)$ and $c(\gamma)$ are analytic, and depend only on the three parameters H, C_1, and α. The knowledge (either by measurements or from theoretical considerations) of these parameters is then enough to compute all the statistical properties of the field. The implicit hypothesis is that this field results from a universal process, hence these parameters are universal. The first, H, is often known theoretically and experimentally, and is therefore already recognized to be universal for many fields. The second, C_1, may perhaps fluctuate slightly with time and location (e.g. Tessier et al., 1993). In fact the most important parameter, the Levy index α, which is fundamental for the classification of the fields (see Tables 1 and 2) is the most likely to be universal. Some experimental results tend to confirm this assumption: at least for the temporal rain rate: five different experiments (Hubert et al., 1993) have (independently) estimated the different time periods, geographical locations, and for both rain gauge accumulations and radar measurements the value $\alpha = 0.5 \pm 0.05$ (see also Lovejoy & Schertzer, 1991, 1992, 1995).

2.6 Scaling anisotropy and generalized scale invariance (GSI)

The standard picture of atmospheric dynamics is that of an isotropic 2-D large scale and an isotropic 3-D small scale, the two separated by a 'meso-scale gap'. Mounting evidence now suggests that, on the contrary, atmospheric fields, while strongly anisotropic, are nevertheless scale invariant right through the meso-scale. The idea of generalized scale invar-

iance (GSI) is to leave the artificial 2-D/3-D dichotomy and to postulate first scale invariance and then study the (unusual) remaining symmetries.

The specification of GSI requires a generator (G) which can be a nonlinear function (varying from point to point): data sets with very large ranges of scale will be needed, and even then, some simplifying approximations will be necessary. As a result of these difficulties the first empirical tests were studies of the compression (stratification) part of GSI associated with the trace of the generator (the elliptical dimension d_{el}). The studies have specifically avoided the difficult differential rotation problem (see below) by concentrating on the vertical stratification Schertzer & Lovejoy 1983, 1985) who estimated $D_{el} = 23/9 = 2.555\ldots$ for the horizontal wind field, $D_{el} = 2.22 \pm 0.07$ for the vertical stratification of rain, and $D_{el} = 2.5 \pm 0.3$ in space/time for the rainfield (Lovejoy et al., 1987; Lovejoy & Schertzer, 1991).

To go from one scale to another, we only need to specify the scale ratios (see Fig. 14, which shows anisotropic scale invariance). We can here define a (semi) group of scale changing operators $T_\lambda = \lambda^{-G}$ (G being the generator) which reduces the scale of vectors by the scale ratio λ: $B_\lambda = T_\lambda(B_1)$ is the ball of all vectors at scale λ (where the unit 'ball' B_1 defines all the unit vectors). Virtually the only other restriction on T_λ is that the B_λ are strictly decreasing ($B_\lambda \supset B_{\lambda'}$; $\lambda < \lambda'$), hence that the real parts of the (generalized) eigenvalues of G are all > 0.

Approximating G by a matrix leads to linear GSI: when there are no off-diagonal elements we obtain only differential stratification, 'self-affine' (multi-)fractals. Off-diagonal elements are associated with differential rotation and can be empirically estimated on scanned cloud satellite images with the help of the Monte Carlo Differential Rotation Technique (Pflug et al., 1993) or (better) by the Scale Invariant Generator Technique (Lewis et al., 1995).

Table 1. *Values of the universal multifractal parameters as estimated by different authors on different data sets.*

	Gauge, daily accumulation	Gauge, 6 minutes resolution	Gauge, daily accumulation	Gauge, daily accumulation	Gauge, 15 minutes resolution
Location	Global network	Réunion Islands (France)	Nîmes (France)	Dédougou (Burkina Faso)	Alps (France)
Stations	4000	1	1	1	28
Duration	1 year (scaling regime up to 16 days)	1 year (scaling regime up to 30 days)	40 years (scaling regime up to 16 days)	45 years	4 years
α	0.5	0.5	0.45	0.59	0.50
C_1	0.6	0.20	0.6	0.32	0.47
Reference		Hubert et al. (1993)	Ladoy et al. (1993)	Hubert et al. (1993)	Desurosne et al. (1995)

Table 2. *The estimated universal multifractal parameters for each group of satellite pictures studied in Tessier et al. (1993). The accuracy on the values of the parameters is about ± 0.1.*

Satellite	Sensor	Wavelength	Scaling range	α	C_1	H
NOAA 9	AVHRR channel 1	0.5 to 0.7 μm	1 to 512 km	1.13	0.09	0.4
NOAA 9	AVHRR channel 2	0.7 to 1.0 μm	1 to 512 km	1.10	0.09	0.4
NOAA 9	AVHRR channel 3	3.6 to 3.9 μm	1 to 512 km	1.11	0.07	0.3
NOAA 9	AVHRR channel 4	10.4 to 11.1 μm	1 to 512 km	1.35	0.10	0.5
NOAA 9	AVHRR channel 5	11.5 to 12.2 μm	1 to 512 km	1.35	0.10	0.5
METEOSAT	VIS	0.4 to 1.1 μm	8 km to 4000 km	1.35	0.10	0.3
METEOSAT	IR	10.5 to 12.5 μm	8 km to 4000 km	1.21	0.09	0.4
LANDSAT	MSS	0.49 to 0.61 μm	166 m to 83 km	1.23	0.07	0.4

3. BEYOND SCALAR MODELING

3.1 Motivations

Until now the multifractal modeling of rain has relied on the simplifying hypothesis that the interaction between rain and the dynamics can be reduced to a scalar relationship (namely between their respective fluxes). This is fundamentally the reason why until now, multifractal results have always been expressed in terms of scalar fields. Theoretically, however, even in the simplest case of passive advection this relation is vectorial (the velocity field coupled with the concentration field via the gradient of the latter[4]). This situation is in a way paradoxical: classical methods, such as those used in GCM modeling, deal easily with this vectorial interaction but on a very limited range of scales, whereas scaling models deal easily with an infinite range of scales but avoid treating this vectorial interaction.

Below, we develop a rather general framework of 'Lie cascades' in order to analyze and generate multiplicative processes for vectorial and tensorial fields. More generally we study the rather abstract fields admitting a Lie group of symmetries. This framework opens a scaling *and* vectorial alternative to GCM techniques, since then we may consider the generator of the joint field $(v, R, I, \ldots)$, (= velocity, rain rate, radiance, etc.) which generates not only each component field, but also their (vectorial, tensorial, etc.) interrelations.

Are the scalar cascade processes in fact restricted to positive scalar fields? If such was the case, then their relevance to turbulence could be quite questionable. Indeed, the energy flux density ε_λ (from larger to smaller scales L/l, $\lambda = [1, \infty]$) in turbulence is not always positive. Using analytical closure schemes (Lesieur & Schertzer, 1976), or the Renormalization Group approach (Forster et al., 1977; Herring et al., 1982), the essential backwards contribution of the flux has been shown to result in a 'beating term' or 'renormalized forces' due to nonlocal interactions. Turbu-

---

[4] And not by a scalar relationship between their respective fluxes, as simplified in multifractal scalar cloud modeling (Wilson et al., 1991; Pecknold et al., 1993).

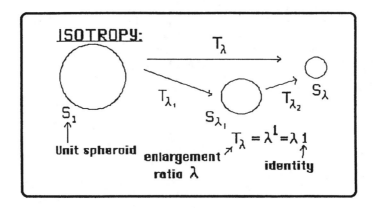

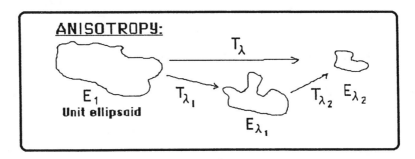

Fig. 14 An illustration of a self-invariant operator giving isotropy as compared to a generalized scale-invariant operator giving anisotropy.

lence clearly cannot be reduced to an eddy-viscosity and a one-way downward energy flux, the balance is somewhat more subtle.

Another fundamental reason to look for extensions of cascade processes is that turbulence is fundamentally vectorial as recalled by Kraichnan (1958). Even more generally mechanics is at least tensorial! This is clearly seen by considering the fundamental continua mechanics equation relating the acceleration a to the stress-tensor σ and the field of forces F:

$$\rho a = \mathrm{div}(\sigma) + F \tag{15}$$

The relationship between the strain tensor D and σ generally involves the fourth order tensor of elasticity. And if we want to investigate the similarities and differences between (fluid) turbulence and seismicity understood as the 'turbulence of solids' (Kagan, 1992), tensorial cascades will be indispensable.

Before solving the problem, let us point out the difficulty. The main problem with a real cascade with alternating signs is that the set of real numbers is not algebraically closed, i.e. it doesn't satisfy d'Alembert's theorem; in particular positive numbers have 2 square roots, negatives none. A rule related to this is the sign of products: products of the same sign give positive numbers, while products of two opposite signs give

negative results. As a consequence there are obviously some nontrivial problems in renormalizing a discrete real cascade by a factor 2 and conversely to introducing intermediate scales. More fundamentally, and especially when one considers a continuous process, a series of multiplications corresponds to an exponentiation of a sum, unfortunately an exponential of any real number is a positive number!

3.2 Complexification of a cascade as an example

By considering the algebraic closure of the real numbers (i.e. the set C of the complex numbers) we should already be able to solve the above mentioned difficulties. For instance, the image of C under exponentiation is C itself. On the other hand, complex multiplication (with $v(=x_1+ix_2)$) corresponds to a particular linear transformation on $\mathcal{R}^2$, i.e. the conformal transformation which is a particular subgroup of $L(\mathcal{R}^2, \mathcal{R}^2)$ (the set of all linear transformations on $\mathcal{R}^2$) which can be identified with the product of rotation (angle θ) and dilation (ratio r) ($v = re^{i\theta}$).

$$v_\lambda = \exp(\Gamma_\lambda)v_1 \quad where \quad v_\lambda, v_1 \in C; \Gamma_\lambda \in C \tag{16}$$

The significance of $\Gamma_{R,\lambda} = \mathrm{Re}(\Gamma_\lambda)$ and $\Gamma_{I,\lambda} = \mathrm{Im}(\Gamma_\lambda)$ is obvious: $\Gamma_{R,\lambda}$ generates a nonnegative cascade process which modulates the amplitude of the modulus of v_λ, whereas $\Gamma_{I,\lambda}$

gives the rotation of v_λ, hence the sign of its real part. We may focus ourselves on the special case where $\Gamma_{R,\lambda}$ and $i\Gamma_{I,\lambda}$ are independent stochastic processes with corresponding characteristic functions $K_R(q)$, $K_I(q)$:

$$\langle v_\lambda^q \rangle = \langle \exp(q\Gamma_{R,\lambda}) \rangle \langle \exp(iq\Gamma_{I,\lambda}) \rangle \langle v_1^q \rangle = \lambda^{K_R(q)} \lambda^{K_I(q)} = \lambda^{K(q)}$$
(17)

The characteristic function $K(q)$ of the complex process is therefore simply:

$$K(q) = K_R(q) + K_I(q) \qquad (18)$$

It is important to note that whereas $K_R(q)$ is real for any real q, $K_I(q)$ is complex, being in general neither real nor pure imaginary. The condition of conservation ($\langle v_\lambda \rangle = 1$) still corresponds to $K(1) = 0$, but not to $K_I(1) = 0$, i.e. Γ_R generates a nonconservative process for the vector modulus. Let us consider as an example (Brethenoux et al., 1992) and as an illustration the discrete lognormal case (Gaussian generator; λ_0 being the fixed step scale ratio). The real and imaginary exponential increments $\Gamma_{R,\lambda,0}$ and $\Gamma_{I,\lambda,0}$ respectively will be a Gaussian variable of variance σ_R^2 and mean m_R (resp. σ_I^2 and m_I) which lead to a generalization of the scalar universal scaling function (eq. (13) with $\alpha = 2$):

$$K_R(q) = C_{1,R}(q^2 - q) - H_R q; \quad C_{1,R} = \sigma_R^2/2; \quad H_R = C_{1,R} - m_R$$
$$K_I(q) = -C_{1,I}(q^2 - q) - H_I q; \quad C_{1,I} = \sigma_I^2/2; \quad H_I = C_{1,I} - im_I$$
$$K(q) = C_1(q^2 - q) - Hq; \quad C_1 = C_{1,R} - C_{1,I}; \quad H = H_R + H_I$$
(19)

A conservative field is obtained with $m_R = -C_1$ (i.e. $\neq -C_{1,R}$, as required to obtain a conservative cascade of modulus), $m_I = 0$. Fig. 15(a)–(e) gives the first steps of the corresponding complex cascade.

One may note that $K(q)$ remains of the standard universal form even for complex q. Similar properties hold for Lévy processes when $\Gamma_{R,\lambda}$ and $\Gamma_{I,\lambda}$ are independently identically distributed. However, $\Gamma_{R,\lambda}$ and $\Gamma_{I,\lambda}$ do not necessarily need to have the same α and there is no longer the requirement that $\Gamma_{I,\lambda}$ should correspond to an extremal Lévy process, since $K_I(q)$ for real q is the Fourier characteristic function of $\Gamma_{I,\lambda}$ whereas $K_R(q)$ remains the Laplace characteristic function of $\Gamma_{R,\lambda}$, and admits the usual scalar universal form (eq. (13), with respectively H_R, $C_{1,R}$, α_R instead of H, C_1, α). The rather more general universal form[5] of $K_I(q)$ is defined for all q (the $\pm$ is the sign of q). Note that β is the asymmetry parameter of the Lévy process $\Gamma_{I,\lambda}$ and $\beta = -1$ for an extremal Lévy process such as $\Gamma_{R,\lambda}$:

$$K_I(q) + H_I q = -\frac{C_{\pm 1,I}}{\alpha - 1}(q|^{\alpha I} - a) \qquad (\alpha \neq 1)$$

$$K_I(q) + H_I q = -C_{\pm 1,I} q |Logq| \qquad (\alpha = 1) \quad (20)$$
$$C_{\pm 1,I} = C_I\{\cos(\pi\alpha/2) \pm i\beta \sin(\pi\alpha/2)\}\Gamma(3 - \alpha_I)\alpha_I \quad (\pm = \text{sgn}(q))$$

with C_I being the coefficient of the canonical Lévy measure dF (cf. eq. (14)) defining $\Gamma_{I,\lambda}$. Fig. 16 displays the complex scaling analysis for a visible and infrared satellite image pair ($v = I_V + iI_R$, I_V and I_R being the visible and infrared radiances respectively).

3.3 Vectorial processes

In the previous subsection, we extended scalar cascades to two component vector cascades by complexifying the cascade. More generally, we may consider nonpositive cascades as being components of more or less straightforward vectorial extensions of positive real processes:

$$v_\lambda = \exp(\Gamma_\lambda)v_1; \qquad v_\lambda, v_1 \in \mathcal{R}^d; \Gamma_\lambda \in L(\mathcal{R}^d, \mathcal{R}^d) \quad (21)$$

the vs being vectorial fields from $\mathcal{R}^d$ to $\mathcal{R}^d$, v_1 being a homogeneous vectorial field (e.g. in the strictest sense: $\forall y \in \mathcal{R}^d v_1(x + y) = v_1(x)$). Just as in the positive scalar case, in order to obtain multiple scaling of the moments Γ_λ should be some band limited $1/f$ noise although now we have a tensor scaling function $K(q)$:

$$\forall \lambda > 1 : \langle \exp(q\Gamma_\lambda) \rangle \approx \lambda^{K(q)}; \quad K(q) \in L(\mathcal{R}^d, \mathcal{R}^d) \quad (22)$$

Introducing furthermore the vectorial singularities γ and their codimensions $c(\gamma)$:

$$\forall \gamma \in \mathcal{R}^d, S_\lambda(\gamma) = \{v \in \mathcal{R}^d, v_i \geq \lambda^{\gamma_i}:\} \qquad Pr(v_\lambda \in S_\lambda(\gamma)) \approx \lambda^{-c(\gamma)} \quad (23)$$

For conservative processes, we still have the same type of conservation law[6]:

$$\langle v_\lambda \rangle = \langle v_1 \rangle; \quad i.e. \quad K(1) = 0 \qquad (24)$$

3.4 Lie groups and their Lie algebra of generators

In fact, *independently* of the representation of the v field and of the B_λ balls (as discussed above), we are only using the (multiplicative) group properties related to the basic fact that scale ratios simply multiply. Using $l = l_1 l_2$ we obtain:

$$T_\lambda = T_{\lambda_1} T_{\lambda_2}; \qquad \tau_\lambda = \tau_{\lambda_1} \tau_{\lambda_2} \qquad (25)$$

T_λ, τ_λ prescribe the *group transformations* respectively of the space transformation acting on the balls B_λ and of the cascade process acting on the fields v_λ:

$$B_\lambda = T_\lambda B_1; \qquad v_\lambda = \tau_\lambda v_1 \qquad (26)$$

Such one-parameter groups can be obtained as the result of stochastic flows obtained from stochastic integrations

---

[5] Which can be obtained with the help of Appendix A of Schertzer & Lovejoy (1991).

[6] Which is a consequence of the martingale property of the process (the conditional expectation at resolution scale λ of $v_{\lambda'}$ ($\lambda' > \lambda$) is v_λ) (see Kahane, 1985).

(a)

(b)

(c)

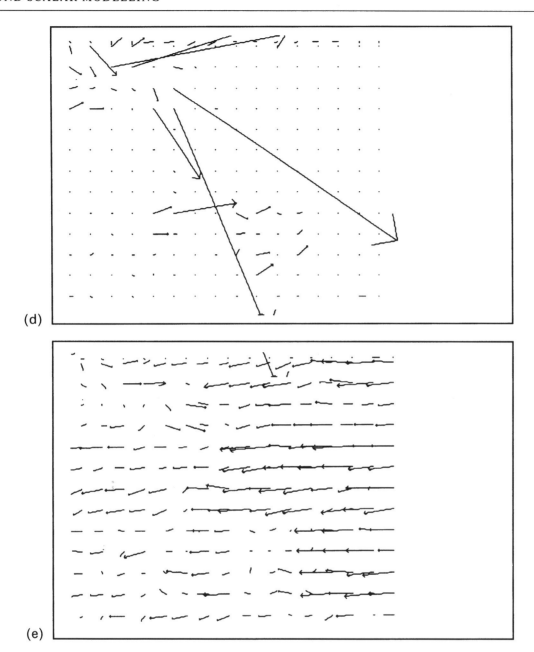

(d)

(e)

Fig. 15 An example of a complex cascade with a Gaussian generator ($\alpha = 2$). (a) Initial homogeneous complex field, (b) the bare field obtained at the first step, (c) the corresponding complex singularities, (d) the bare field obtained at the third step, (e) the corresponding complex singularities.

(more precisely from Stratatovich integrations as discussed[7] in Schertzer & Lovejoy 1995) over infinitesimal (random) generators $d\Gamma$ and dG:

$$d\varepsilon_\lambda = \varepsilon_\lambda d\Gamma_\lambda; \quad dT_\lambda = T_\lambda dG_\lambda \quad (27)$$

Originally (Schertzer & Lovejoy, 1991) such an integration was proposed only on T_λ in the case of the so-called nonlinear

(random) GSI (generalized scale invariance). The solutions of the above equations will be denoted as e^{Γ_λ} and e^{G_λ} respectively.

Corresponding to the group property of the transformation of the field or of the space, there is a Lie algebra structure for the generators, i.e. there is a skew and distributive product [,], called the Lie bracket. The group properties of the statistical moments of the field or scale transformations correspond to the fact that the second characteristic function (cumulant generating function) $K_\lambda(q)$ generates (for the

---

[7] Contrary to the most popular stochastic integration, i.e. the Ito integration, the Stratatovich integration corresponds to a centred integration (e.g. Kunita, 1990).

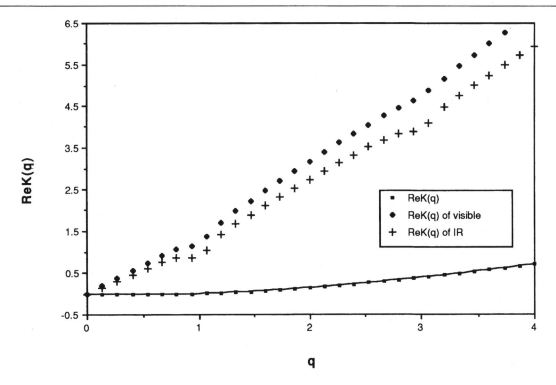

Fig. 16 The moment scaling functions are shown for a visible and infrared satellite image pair ($K_V(q)$, $K_{IR}(q)$ respectively) at 8 km resolution taken over the Montreal area as part of the RAINSAT automated satellite rain algorithm (each image is 256×256 pixels). Also shown is the $K_R(q)$ function described in the text which is obtained by considering the multiscaling of the modulus of the vector (visible, infrared). This gives partial information about the scaling interrelation between the two fields. Using the double trace moment (Lavallée, 1991; Lavallée et al., 1991, 1993), we obtain $\alpha_V = 1.7$, $\alpha_{IR} = 1.75$, $\alpha_R = 1.73$, $C_{1,V} = 0.22$, $C_{1,IR} = 0.20$, $C_{1,R} = 0.25$ (accuracy of the estimates is about ± 0.2).

different values of q) a Lie subalgebra, e.g. for the v field:

$$d\langle v_{\lambda}^{q} \rangle = \langle v_{\lambda}^{q} \rangle dK_{\lambda}(q) \quad \text{or} \quad \langle v_{\lambda}^{q} \rangle = e^{K_{\lambda}(q)} \qquad (28)$$

We are now able to give a precise notion of scaling:

$$dK_{\lambda} \approx K d\lambda/\lambda \quad \text{or} \quad e^{K_{\lambda}} \approx \lambda^{K} \qquad (29)$$

This already shows that the characteristic function K_{λ} should be log divergent in the scale ratio λ just as in the framework of the scalar cascade. It can further be shown (Schertzer & Lovejoy, 1993) that the log divergence of K_{λ} still corresponds to the fact that the generator is a 'pink noise', i.e. having a (generalized) spectrum being exactly inverse to the wave number.

3.5 A quaternion-like representation of $L(\mathscr{R}^2, \mathscr{R}^2)$ as a preliminary example

Considering the linear transformations of the plane, we need not restrict our attention to conformal ones, which correspond to complex multiplications. Indeed, one basis $\{Y_i\}$ of the matrix representation of $L(\mathscr{R}^2, \mathscr{R}^2)$ is given by the 4 following matrices:

$$1 = Y_1 = \begin{bmatrix} 1 & 0 \\ 0 & 1 \end{bmatrix}; \quad I = Y_2 = \begin{bmatrix} 0 & -1 \\ 1 & 0 \end{bmatrix};$$

$$J = Y_3 = \begin{bmatrix} 0 & 1 \\ 1 & 0 \end{bmatrix}; \quad K = Y_4 = \begin{bmatrix} 1 & 0 \\ 0 & -1 \end{bmatrix}; \qquad (30)$$

which satisfy the following anti-commutation relations (the anticommutator is defined as $\{A, B\} = (AB + BA)$:

$$\{Y_2, Y_3\} = 0, \quad \{Y_2, Y_4\} = 0, \quad \{Y_3, Y_4\} = 0 \qquad (31)$$

Compared to the complexification example (Section 3.2) this has a richer structure due to the addition of the 'parity' operator J and the 'conjugation' operator K. With real and independent pink noises Γ^i, $\Gamma = \Gamma^i Y_i$ is a scaling tensor field (satisfying eq. 29). Its characteristic function $K_{\Gamma}(q)$ will be simply the sum of the characteristic functions $K_i(q)$ of the Γ^i. Below we see that when quaternions and Clifford algebra are considered an even richer structure is obtained.

3.6 Classification and factorization of the Lie processes

Bearing in mind the very general Lie framework, we now can study a system which is invariant under different symmetries, not only scaling symmetries. We are naturally led to look for a kind of classification of the possible algebra. The answer is rather classical and half-positive: any Lie algebra can be decomposed[8] into a (semi-direct[9]) sum into a 'semi-simple algebra' σ and its radical R. Whereas there exists a universal Cartan classification (e.g. Sattinger & Weaver, 1985) of the (real or complex) semi-simple algebra (e.g. the celebrated $so(n)$...), no such result exists for radicals, which not only are totally different from semi-simple algebras, but correspond, as pointed out below, to scale symmetries. It means we are entering a particular world of symmetries which has not yet been explored. Indeed, whereas a *semi-simple* algebra is one having no abelian ideals[10] (other than 0), the radical R contains abelian ideals being defined as the maximal solvable ideal[11]. A trivial but fundamental example of an abelian ideal is the ideal generated by the identity $\{\lambda 1 | \lambda \in \mathcal{R} \text{ or } \in C\}$, hence the trivial scaling, it corresponds to the stronger property of nilpotency[12].

On the one hand, the solvability of the radical prevents it respecting a simple universal classification, on the other hand, it yields a simple generalization of factorization. Indeed, for a generator defined on an abelian ideal s, then the field merely factorizes as a product of fields generated by a element of the basis X_i ($\Gamma = \Gamma^i X_i$) of s ($[X_1, X_2, \ldots, X_n]$ spanning s):

$$\varepsilon_\lambda = \prod_{i=1}^{n} \lambda^{\Gamma^i X_i} \tag{32}$$

where the Γ_i are pink noises. Such a factorization can be extended to the whole radical of the algebra thanks to the Lie theorem on solvable algebra and a Yamato-Kunita theorem (Kunita, 1990). Indeed, the factorization still holds in the following sense: $\{X_1, X_2, \ldots, X_i\}$ which spans the increasing ideals $g_i([R, g_{i-1}] = g_i; \; R = g_n \supset g_{n-1} \supset \ldots \supset g_1 \supset 0, \; \dim(g_i) = i.)$, then the Γ^i are replaced in eq. (32) by N^i which are sums, or products, or integrals, or exponentials of the Γ^j. In the case of nilpotency[13], which may be relevant for scaling, exponentials do not intervene, hence we are back to a rather simple factorization.

3.7 Quaternions and Clifford algebra as examples

Quaternions (for the dimension 4) and Clifford algebra C_n (for the dimensions $2^n, n > 2$) are the real linear Lie algebra (a subalgebra of $L(C^n, C^n)$) defined by the following anti-commutators relations corresponding to a (Pauli) factorization of the Laplacian[14]:

$$\{\alpha^\mu, \alpha^\nu\} = 2\delta^{\mu\nu} \tag{33}$$

This algebra is generated by $1, \alpha_1, \ldots, \alpha_n$ and all their products $\alpha_i.\alpha_j\ldots\alpha_l$ (which can be always be ordered, since $\alpha_i.\alpha_j = -\alpha_j.\alpha_i, \alpha_i^2 = 1$), thus contains 2^n elements. For the quaternions:

$$\alpha_1 = \begin{bmatrix} 0 & 1 \\ 1 & 0 \end{bmatrix}; \quad \alpha_2 = \begin{bmatrix} 0 & i \\ -i & 0 \end{bmatrix} \tag{34}$$

and they can be expressed with the help of the Pauli matrices σ_i:

$$\sigma_1 = \alpha_1; \quad \sigma_2 = -\alpha_2; \quad \sigma_3 = -\alpha_1.\alpha_2 = \begin{bmatrix} i & 0 \\ 0 & -i \end{bmatrix} \tag{35}$$

For $n = 4$, we have:

$$\alpha_1 = \begin{bmatrix} \sigma_1 & 0 \\ 0 & \sigma_1 \end{bmatrix}; \quad \alpha_2 = \begin{bmatrix} \sigma_1 & 0 \\ 0 & \sigma_2 \end{bmatrix}; \quad \alpha_3 = \begin{bmatrix} 0 & \sigma_3 \\ \sigma_3 & 0 \end{bmatrix}; \quad \alpha_4 = \begin{bmatrix} 0 & -i\sigma_3 \\ i\sigma_3 & 0 \end{bmatrix} \tag{36}$$

More generally $n = 2m$ or $2m + 1$, and for any $j \leq m$; the α_j can be expressed on $(C^2)^{\otimes n}$ (the spinors space) and for $n = 2m + 1$; $\alpha_{2m+1} = \sigma_3^{\otimes 2m+1}$. Not only do these well-known examples allow us to generalize rather straightforwardly the result of complex cascades to $n > 1$ (simulations as well as data analysis, e.g. multichannel radiance field will be developed in subsequent works) but also, as mentioned earlier they correspond in a given representation to well-defined equations. This shows that the Lie algebra of the generators of symmetries might well be the indispensable tool necessary to bridge the gap conclusively between stochastic models and deterministic equations.

4. CONCLUSIONS

In Schertzer & Lovejoy (1987), it was proposed that in many geophysical applications scaling symmetries can be used as dynamical constraints instead of coupled nonlinear partial differential equations. This is the usual physics notion that a system is totally determined once all its symmetries are known. Since scaling arguments are so general it has become urgent to develop a formalism for handling scaling for coupled multifractal processes (e.g. vector cascades) as well as for restricting the generator of the scaling symmetries using additional symmetries. Both goals require the formalism of the Lie cascades developed here.

---

[8] Via a Levi decomposition.
[9] The sum of of two subalgebras a and b is direct when the two commute ($[a, b] = 0$), it is semi-direct when $a \supset [a, b]$.
[10] A subalgebra s is an ideal of g if $s \supset [s, g]$ (i.e. not only $[s, s]$).
[11] I.e. the largest ideal leading by a nested sequence of ideals to an abelian ideal.
[12] The limit of a nested sequence of ideals is not only an abelian ideal, but it commutes with the whole algebra.
[13] Nilpotency is a slightly stronger property than solvability.

[14] $(\alpha^\mu \partial_\mu)^2 = \partial_\mu^2$.

We have not only shown there is no fundamental reason to restrict cascade processes to positive (real) quantities, but there are also very wide possible generalizations to rather abstract processes. As a consequence, a potentially wider unity of geophysics is restated in a new way. At the same time a quantitative way to classify the wide diversity of phenomena which occur on a large range of scales is pointed out with the help of the classification of the corresponding Lie algebra. This classification will also enable us to discover new types of nonlinear interactions.

Immediate applications of the ideas discussed here include the simulation of vector multifractals and the scale invariant characterization of the interrelations of rain, cloud radiance and other fields. We gave a first example of the latter by analyzing the multiscaling of the vector moments of the joint visible and infrared cloud radiance fields from GOES satellite data. When this is extended to radar reflectivities of rain and rain gauge measurements, the resolution independent characterization of their interrelation can form the basis of resolution independent satellite rain algorithms, as well as for the resolution independent calibration of radars from rain gauges.

ACKNOWLEDGMENTS

We acknowledge stimulating discussions with Y. Chigirinskaya, P. Hubert, Y. Kagan, D. Lavallée, D. Marsan, C. Naud, G. Salvadori, F. Schmitt and Y. Tessier. We thank G. Brethenoux, D. Mitrani and J. Dezani for help with the simulation of a complex cascade, F. Francis and C. Lugania for help with the preliminary complex analyses and for the preparation of the corresponding figures.

REFERENCES

AUSTIN, P.M. & R.A. HOUZE, 1972: Analysis of structure of precipitation patterns in New England. J. Appl. Meteor., 11, 926–935.

BAK, P., C. TANG & K. WIESSENFELD, 1987: Self-Organized Criticality: an explanation of 1/f noise. Phys. Rev. Lett. 59, 381.

BENZI, R., G. PALADIN, G. PARISI & A. VULPIANI, 1984: On the multifractal nature of fully developed turbulence. J. Phys. A, 17, 3521–3531.

BIALAS, A. & R. PESCHANSKI, 1986: Moments of rapidity distributions as a measure of short-range fluctuations in high-energy collisions, Nucl. Phys. B, B 273, 703–718.

BRETHENOUX, G., D. MITRANI, J. DEZANI, D. SCHERTZER & S. LOVEJOY, 1992: Lie cascades: multifractal vectorial and tensorial multipicative processes. EOS, 73, 14 supp., 57.

CHIGIRINSKAYA, Y., D. SCHERTZER, S. LOVEJOY, A. LAZAREV & A. ORDANICH, 1994: Unified multifractal atmosphere dynamic tested in the tropics: part I, Horizontal scaling and self-criticality. Nonlinear Processes in Geophysics, I, 105–114.

DESUROSNE, I., P. HUBERT & G. OBERLIN, 1995: Analyses fractales et multifractales: Une étape vers la formulation d' un modèle prédéterministe des précipitations d'altitude. prepared for Hydro. Continent.

FAN, A.H., 1989: Chaos additif et multiplicatif de Lévy. C. R. Acad. Sci. Paris I, 308, 151.

FELLER, W. 1971: An introduction to probability theory and its applications, vol.2, Wiley, New York.

FORSTER, D., D.R. NELSON & M.J. STEPHENS, 1977: Large distance and long time properties of a randomly stirred field. Phys. Rev., A16, 732–749.

FRISCH, U., P.L. SULEM & M. NELKIN, 1978: A simple dynamical model of intermittency in fully developed turbulence. J. Fluid Mech., 87, 719.

GRASSBERGER, P., 1983: Generalized dimensions of strange attractors. Phys. Lett., A 97, 227.

HALSEY, T.C., M.H. JENSEN, L.P. KADANOFF, I. PROCACCIA & B. SHRAIMAN, 1986: Fractal measures and their singularities: the characterization of strange sets. Phys. Rev. A., 3 1141.

HENTSCHEL, H.G.E. & I. PROCCACIA, 1983: The infinite number of generalized dimensions of fractals and strange attractors. Physica, 8D, 435.

HERRING, J.R., D. SCHERTZER, M. LESIEUR, G.R. NEWMAN, J.P. CHOLLET & M. LARCHEVEQUE, 1982: A comparative assessment of spectral closures as applied to passive scalar diffusion. J. Fluid Mech., 124, 411–437.

HUBERT, P. & J.P. CARBONNEL, 1989: Dimensions fractales de l'occurrence de pluie en climat soudano-sahélien. Hydrol. continent., 4, 3–10.

HUBERT, P., Y. TESSIER, S. LOVEJOY, D. SCHERTZER, P. LADOY, J. P. CARBONNEL & S. VIOLETTE, 1993: Multifractals and extreme rainfall events. Geophys. Res. Lett., 20, 10, 991–934.

KAGAN, Y.Y., 1992: Seismicity turbulence of solids, Nonlinear Sci Today, 2, 1–13.

KAHANE, J.P., 1985: Sur le chaos multiplicatif, Ann. Sci. Math. Que., 9, 435.

KOLMOGOROV, A.N., 1941: Local structure of turbulence in an incompressible liquid for very large Reynolds numbers. Dokl. Acad. Sci. USSR. 30,299.

KOLMOGOROV, A.N., 1962: A refinement of previous hypotheses concerning the local structure of turbulence in viscous incompressible fluid at high Reynolds number. J. Fluid Mech., 83, 349.

KRAICHNAN, K.R., 1958: Irreversible statistical mechanics of incompressible hydrodynamic turbulence. Phys. Rev., 109, 1407–1422.

LADOY, P., LOVEJOY, S. & D. SCHERTZER, 1991: Extreme fluctuations and intermittency in climatological temperatures and precipitation, in: Scaling, fractals and non-linear variability in geophysics, D. Schertzer &, S. Lovejoy eds., 241–250, Kluwer, Dordrecht.

LADOY, P., F. SCHMITT, D. SCHERTZER & S. LOVEJOY, 1993: Variabilité temporelle multifractale des observations pluviométriques à Nîmes. C. R. Acad. Sci. Paris, II, 317, 775–782.

LAVALLÉE, 1991: Multifractal analysis and simulation techniques and turbulent fields, Ph.D. Thesis, McGill University, Montréal, 132 (1991).

LAVALLÉE, D., D. SCHERTZER & S. LOVEJOY, 1991: On the determination of the co-dimension function. Scaling, fractals and non-linear variability in geophysics, D. Schertzer & S. Lovejoy eds., 99–110, Kluwer, Dordrecht.

LAVALLÉE, D., S. LOVEJOY, D. SCHERTZER & P. LADOY, 1993: Nonlinear variability, multifractal analysis and simulation of landscape topography, in Fractals in Geography, L. De Cola & N. Lam eds., Kluwer, 158–192, Dordrecht-Boston.

LAVALLÉE, D., S. LOVEJOY, D. SCHERTZER & F. SCHMITT, 1992: On the determination of universal multifractal parameters in turbulence. Topological aspects of the dynamics of fluids and plasmas, Eds. K. Moffat, M. Tabor & G. Zaslavsky, p.463–478, Kluwer, Dordrecht.

LAZAREV, A., D. SCHERTZER, S. LOVEJOY & Y. CHIGIRINSKAYA, 1994: Unified multifractal atmospheric dynamics tested in the tropics, part II, Vertical scaling and generalized scale invariance. Nonlinear Processes in Geophysics, I, 115–123.

LESIEUR, M. & D. SCHERTZER, 1978: Amortissement autosimilaire d'une turbulence à grand nombre de Reynolds. J. Méc., 17, 607–646.

LEVICH, E. & E. TZVETKOV, 1985: Helical inverse cascade in three-dimensional turbulence as a fundamental dominant mechanism in meso-scale atmospheric phenomena, Phys. Rep., 128, 1–37.

LEVITCH, E. & I. SHTILMAN, 1991: Helicity fluctuations and coherence in developed turbulence. Nonlinear Variability in Geophysics: Scaling and Fractals, Eds. D. Schertzer & S. Lovejoy, Kluwer, Dordrecht, 13–30.

LEWIS, G., S. LOVEJOY & D. SCHERTZER, 1995: The scale invariant generator technique for parameter estimates in generalized scale invariant systems. Submitted to Nonlinear Processes in Geophysics.

LOVEJOY, S. 1981: Analysis of rain areas in terms of fractals, 20th conf. on radar meteorology, 476–484, AMS Boston.

LOVEJOY, S. & B. MANDELBROT, 1985: Fractal properties of rain and a fractal model. Tellus, 37A, 209–232.

LOVEJOY, S. & D. SCHERTZER, 1985: Generalised scale invariance and fractal models of rain, Wat. Resour. Res., 21, 1233–1250.

LOVEJOY, S., D. SCHERTZER & A.A. TSONIS, 1987: Functional box-counting and multiple elliptical dimensions in rain. Science, 235, 1036.

LOVEJOY, S. & D. SCHERTZER, 1990: Multifractals, universality classes and satellite and radar measurements of cloud and rain fields, J. Geophy. Res., 95, 2021.

LOVEJOY S. & D. SCHERTZER, 1991: Multifractal analysis techniques and the rain and cloud fields from 10^{-3} to 10^6m.. Nonlinear Variability in Geophysics: Scaling and Fractals, D. Schertzer & S. Lovejoy Eds, Kluwer, Dordrecht, 111–144.

LOVEJOY, S. & D. SCHERTZER, 1995: Multifractals and rain. In New uncertainty concepts in hydrology and hydrological modelling, Ed. A. W. Kundzewicz, Cambridge University Press, in press.

LOVEJOY, S., D. SCHERTZER, P. SILAS, Y. TESSIER & D. LAVALLÉE, 1993: The unified scaling model of the atmospheric dynamics and systematic analysis of scale invariance in cloud radiances. Annales Geophysicae, 11, 119–127.

MANDELBROT, B., 1974: Intermittent turbulence in self-similar cascades: Divergence of high moments and dimension of the carrier. J. Fluid Mech., 62, 331.

MANDELBROT, B., 1982: The fractal geometry of nature, W.H. Freeman, New York.

MANDELBROT, B., 1991: Random multifractals: negative dimension and the resulting limitation of the thermodynamic formalism in turbulence and stochastic processes, Eds. J.C.R. Hunt, O.M. Phillips & D. Williams, The Royal Society.

MENEVEAU, C. & K.R. SREENIVASAN, 1987: Simple multifractal cascade model for fully developed turbulence. Phys. Rev. Lett., 59, 13, 1424.

MENEVEAU, C. & K.R. SREENIVASAN, 1989: Measurement of $f(\alpha)$ from scaling of histograms, and applications to dynamical systems and fully developed turbulence. Phys. Lett. A, 137, 3, 103.

NOVIKOV, E.A. & R. STEWART 1964: Intermittency of turbulence and spectrum of fluctuations in energy-dissipation, Izv. Akad. Nauk. SSSR, Ser. Geofiz, 3, 408.

OBUKHOV, A., 1962: Some specific features of atmospheric turbulence. J. Geophys. Res., 67, 3011.

PALADIN, G. & A. VULPIANI, 1987: Anomalous saling laws in multifractals objects. Phys. Rev. Lett., 156, 147.

PARISI, G. & U. FRISCH, 1985: A multifractal model of intermittency. Turbulence and predictability in geophysical fluid dynamics and climate dynamics. Eds. M. Ghil, R, Benzi, G. Parisi, North-Holland, Amsterdam, 84.

PECKNOLD, S., S. LOVEJOY, D. SCHERTZER, C. HOOGE & J.F. MALOUIN, 1993: The simulation of universal multifractals. in: cellular automata: prospects in astronomy and astrophysics, Eds. J.M. Perdang & A. Lejeune, 228–267, World Scientific, Singapore.

PIETRONERO, L., & A.P. SIEBESMA 1986: Self-similarity of fluctuations in random multiplicative processes. Phys. Rev. Lett., 57, 1098.

PFLUG, K., S. LOVEJOY & D. SCHERTZER, 1993: Generalized scale invariance. Differential rotation and cloud texture: analysis and simulation. J. Atmos. Sci., 50, 538–553.

RICHARDSON, L.F., 1922 (republished by Dover, New York, 1965): Weather prediction by numerical process, Cambridge University Press, Cambridge.

SCHERTZER, D. & S. LOVEJOY, 1983: On the dimension of atmospheric motions. Turbulence and chaotic phenomena in fluids, Ed. Tatsumi, Elsevier North-Holland, New York, 505.

SCHERTZER D. & S. LOVEJOY, 1985: The dimension and intermittency of atmospheric dynamics, Turbulent Shear flow 4, Ed. B. Launder, Springer, Berlin, 7.

SCHERTZER, D. & S. LOVEJOY, 1987: Physically based rain and cloud modeling by anisotropic, multiplicative turbulent cascades. J. Geophys. Res. 92, 9693.

SCHERTZER, D. & S. LOVEJOY, 1987b: Singularités anisotropes, divergence des moments en turbulence. Ann. Sc. Math. Que., II, 139–181.

SCHERTZER, D. & S. LOVEJOY, 1989: Nonlinear variability in geophysics: multifractal analysis and simulations. Fractals: Physical origin and consequences, Ed. L. Pietronero, Plenum, New York, 49.

SCHERTZER, D. & S. LOVEJOY, 1991: Nonlinear geodynamical variability: Multiple singularities, universality and observables. Scaling, fractals and non-linear variability in geophysics, Eds. D. Schertzer & S. Lovejoy, Kluwer, Dordrecht, 41–82.

SCHERTZER, D. & S. LOVEJOY, 1992: Hard and Soft Multifractal processes: Physica A, 185, 187–194.

SCHERTZER, D. & S. LOVEJOY, 1994: Multifractal Generation of Self-Organized Criticality. Fractals in the Natural and Applied Sciences, M.M. Novak ed., 325–339, Elsevier, Amsterdam.

SCHERTZER, D. & S. LOVEJOY, 1995: Multifractals and turbulence: fundamental and applications, World Scientific, Singapore, 230 pp. (in press).

SCHERTZER, D., S. LOVEJOY, R. VISVANATHAN, D. LAVAL-LÉE & J. WILSON, 1988: Multifractal analysis techniques and rain and clouds fields. In fractal aspects of materials: disordered systems, Weitz et al. eds, 267–270, Materials Research Society, Boston.

SCHERTZER, D., S. LOVEJOY & D. LAVALLÉE, 1993: Generic multifractalphase transitions and self-organized criticality. Cellular Automata: prospects in astronomy and astrophysics, Eds. J.M. Perdang & A. Lejeune, 216–227, World Scientific, Singapore, in press.

SCHERTZER, D., S. LOVEJOY, D. LAVALLÉE & F. SCHMITT, 1991: Universal hard multifractal turbulence, theory and observations. Nonlinear dynamics of structures. Eds. R.Z. Sagdeev, U. Frisch, F. Hussain, S.S. Moiseev & N.S. Erokhin eds., World Scientific, Singapore, 213–235.

SCHMITT, F., D. LAVALLÉE, D. SCHERTZER & S. LOVEJOY, 1992a: Empirical determination of universal multifractal exponents in turbulent velocity fields. Phys. Rev. Lett., 68, p305–308.

SCHMITT, F., S. LOVEJOY, D. SCHERTZER, D. LAVALLÉE & C. HOOGE, 1992b: Les premières estimations des indices de multifractalité dans le champ de vent et de température. C. R. Acad. Sci. Paris, II, 314, 749–754.

SCHMITT, F., D. SCHERTZER, S. LOVEJOY & Y. BRUNET, 1994: Empirical study of multifractal phase transitions in atmospheric turbulence. Nonlinear Processes in Geophysics, I, 95–104.

SREENIVASAN, K.R. & C. MENEVEAU, 1988. Singularities of the equations of fluid motion. Phys. Rev. A. 38 12, 6287.

TESSIER, Y., S. LOVEJOY & D. SCHERTZER, 1993: Universal multifractals in rain and clouds: theory and observations. J. Appl. Meteor., 32,2, p223–250.

WAYMIRE, E., V.K. GUPTA & I. RODRIGUEZ-ITURBE, 1984: A spectral theory of rainfall intensity at the meso-beta scale, Wat. Resour. Res. 20, 1453–1465.

WILSON, J., S. LOVEJOY & D. SCHERTZER, 1991: Physically based cloud modelling by scaling multiplicative cascade processes. Scaling, fractals and non-linear variability in geophysics, Eds. D. Schertzer & S. Lovejoy, 185–208, Kluwer, Dordrecht.

YAGLOM, A.M., 1966: The influence of the fluctuation in energy dissipation on the shape of turbulent characteristics in the inertial interval, Sov. Phys. Dokl., 2, 26.

Fractals et multifractals appliqués à l'étude de la variabilité temporelle des précipitations

P. HUBERT

URA-CNRS 1367, Ecole des Mines de Paris,

Fontainebleau,

France

ABSTRACT Rainfall exhibits at every time scale a great variability which becomes extreme for short durations. We first tried to give rainfall occurrence a fractal dimension the main interest of which is to be time scale invariant. This geometrical approach appears to be of limited value, the fractal dimension being dependent upon the intensity threshold used to define the rainy character of a given period. This problem can be overcome by substituting multifractal fields to fractal sets.

The fundamental equation of such fields enables us to relate at every scale the fraction of space occupied by singularities to their probability of appearance. This equation depends only on two parameters characterizing respectively departures of the field under study from homogeneity and monofractality. A time scale invariant frequency-intensity-duration formula has been derived within this frame, which suggests the existence for all durations of a possible maximum precipitation.

1. FRACTALS ET MULTIFRACTALS APPLIQUÉS À L'ÉTUDE DE LA VARIABILITÉ TEMPORELLE DES PRÉCIPITATIONS

La pluie est un phénomène qui se manifeste dans l'espace et dans le temps. On peut supposer l'existence d'une fonction $I(x, t)$, caractérisant l'intensité des précipitations au point x de l'espace à deux dimensions constitué par la surface terrestre et au temps t, cette intensité étant exprimée en hauteur d'eau par unité de temps, $[L][T]^{-1}$. Nous ne connaissons a priori rien des propriétés de cette fonction mis à part l'hypothèse de définition en tout point, mais différents types de mesurages permettent d'en estimer des intégrales selon le temps et/ou l'espace (Fig. 1).

Les pluviomètres ou les pluviographes permettent de mesurer quasi ponctuellement des hauteurs de précipitation selon différents pas de temps, allant de la journée à quelques secondes, qui sont des intégrales selon le temps de l'intensité des précipitations en un point.

$$H_{t_1, t_2}(x) = \int_{t_1}^{t_2} I(x, t)dt \tag{1}$$

Une mesure spatiale parfaite (comparable en première approximation à une image satellitaire ou radar) réaliserait une intégration de la fonction I dans l'espace, selon un pixel de surface Σ, à un instant donné, et permettrait d'atteindre la valeur moyenne de l'intensité pluviométrique sur le pixel à cet instant.

$$I_\Sigma(t) = \frac{1}{\Sigma} \iint_\Sigma I(x, t)dx \tag{2}$$

Nous disposons de telles mesures, et elles nous permettent tout d'abord d'apprécier la variabilité et l'intermittence des précipitations, dans le temps et dans l'espace. On l'observe sur n'importe quel hyétogramme, mais on n'en mesure pas toujours le degré extrême. Nous avons étudié de nombreux hyétogrammes recueillis au Niger dans le cadre de l'expérience EPSAT (Lebel et al, 1992); la mesure des durées pendant lesquelles différents seuils d'intensité ont été atteints ou dépassés permet de se rendre compte que la moitié des précipitations annuelles tombe en quelques heures! Sur la Fig. 2, représentant une image radar, on remarquera l'émiettement et la complexité des contours des zones précipitantes.

L'extrême variabilité des précipitations que l'on retrouve à

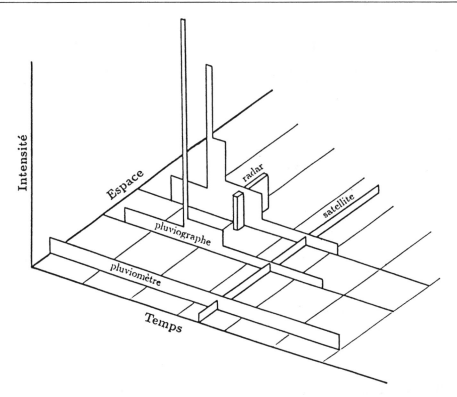

Fig. 1 Les différents appareils de mesure dont nous disposons (pluviomètres, pluviographes, radars, satellites) permettent d'estimer, directement ou indirectement, des intégrales dans le temps et/ou l'espace (réduit à une seule dimension sur cette figure) de l'intensité pluviométrique supposée définie en tout point et à chaque instant.

toutes les échelles de temps et d'espace pose de redoutables problèmes théoriques et pratiques, que l'approche fractale, et surtout multifractale, peut contribuer à aborder et à résoudre (Hubert et Carbonnel, 1988; Lovejoy et Schertzer, 1986, 1991). Nous nous intéresserons ici à quelques aspects de cette approche dans le domaine temporel.

Les fractals ont d'abord été considérés d'un point de vue purement géométrique (Mandelbrot,1975;1982). Sous cet angle, la pluie est considérée comme un ensemble auquel un point (de l'axe du temps) appartient ou n'appartient pas. Nous sommes donc amenés à définir précisément l'occurrence de pluie, ce qui ne peut être réalisé rigoureusement que par rapport à une aire, à un intervalle de temps et à un seuil de précipitation. Un intervalle de temps sera dit pluvieux pour une aire donnée si une quantité d'eau météorique supérieure à un seuil donné a été précipitée sur cette aire pendant l'intervalle de temps considéré. Le pluviomètre permet de définir avec précision une aire d'observation qui se confond avec sa surface de collecte (400 cm^2 en général). Ce type d'appareil est habituellement relevé quotidiennement à 8 heures. On peut alors déterminer l'état de périodes successives de 24 heures, état sec lorsque la hauteur d'eau recueillie est inférieure à un seuil donné (ce seuil étant généralement choisi égal à 0.1 mm), état pluvieux dans le cas contraire.

Nous avons appliqué cette définition à une série de 45

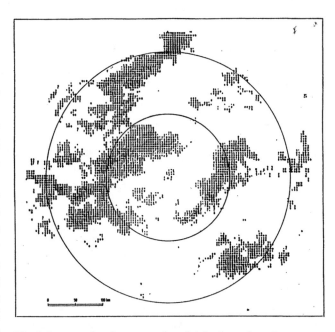

Fig. 2 Image radar d'une zone de précipitations observée sur l'océan Atlantique au large de l'Afrique (expérience GATE). La taille des pixels est de 4 × 4 km, et ils sont représentés d'autant plus foncés que le taux de précipitations est intense (d'après Hudlow et Paterson, 1979; cités par Barett et Martin, 1981).

Fig. 3 45 années d'observations pluviométriques journalières à Dédougou (Burkina Faso). Chaque ligne correspond à une année d'observations. La première ligne, en trait plein, permet d'apprécier la longueur totale de l'année du 1 Janvier au 31 Décembre. Sur les lignes suivantes représentant les années 1922 à 1966 seules les journées pluvieuses au seuil de 0.1 mm ont été soulignées par un tracé (d'après Hubert et Carbonnel, 1989).

années de relevés pluviométriques journaliers de Dédougou (Burkina Faso, 3.39° Longitude Ouest, 12.28° Latitude Nord). L'occurrence de la pluie apparaît alors comme un ensemble disjoint inclus dans l'espace du temps (Fig. 3). Géométriquement, cet ensemble évoque le résultat obtenu après un certain nombre d'itérations dans le processus générateur d'une poussière de Cantor aléatoire à partir d'une barre. Ce type d'objet est devenu classique depuis les travaux de Mandelbrot qui a introduit la notion d'objet fractal et de dimension fractale (dimension de Haussdorf-Besicovitch) dans les sciences de la nature. Cette ressemblance nous a incités à tenter de quantifier en termes dimensionnels l'occurrence de la pluie sur un intervalle de temps (Hubert et Carbonnel, 1989).

Nous avons estimé la dimension fractale de l'occurrence de pluie grâce à la méthode du comptage de boîtes (Hentschel et Proccacia, 1983; Lovejoy, Schertzer et Tsonis, 1987). Etant

donné un objet de dimension fractale D, inclus dans un espace de dimension euclidienne E, si on réalise un maillage de cet espace en boîtes de coté a (selon la valeur de $E = 1, 2$ ou 3, les boîtes seront des segments, des carrés ou des cubes), le nombre N de boîtes nécessaires pour recouvrir l'objet fractal considéré est une fonction de a telle que:

$$Log[N(a)] = -D\,Log[a] \qquad (3)$$

Dans le cas de l'ensemble que nous nous proposons d'étudier, qui est inclus dans un ensemble de dimension 1, les boîtes sont réduites à des segments et la dimension fractale de l'occurrence de pluie sera donc nécessairement comprise entre 0 et 1. $N(a)$ étant déterminé pour différentes valeurs de a, D pourra être estimé comme l'opposé de la pente de la régression de N en a tracée sur un diagramme Log-Log.

Nous avons utilisé pour la station de Dédougou une séquence de $2^{14} = 16\,384$ jours consécutifs (soit près de 45 ans)

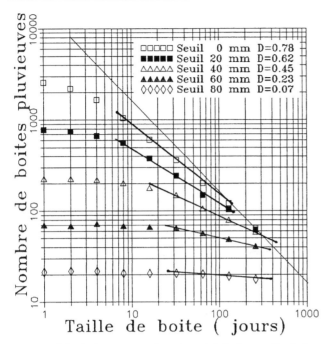

Fig. 4 Méthode du comptage fonctionnel de boîtes appliquée à une série de 16384 hauteurs de pluie journalières de la station de Dédougou (Burkina Faso). Une dimension fractale de l'occurrence de pluie à été estimée pour différents seuils de hauteur de pluie.

débutant le 1 Janvier 1922. Nous avons reporté sur la Fig. 4 (courbe supérieure) les résultats de 10 comptages réalisés pour des maillages de la séquence temporelle considérée selon des segments dont les longueurs sont les termes d'une progression géométrique de premier terme 1 et de raison 2 (1 à 512 jours).

On constate, pour des durées allant de quelques jours à quelques mois, un bon alignement des points représentatifs selon une droite de pente −0.78. Cet alignement confère à l'occurrence de pluie, pour cette plage d'échelles de temps, une structure autosimilaire dont la dimension fractale est égale à 0.78. Cette structure et sa dimension peuvent être rapprochées de la place qu'occupe la saison des pluies dans l'année en Afrique soudano-sahélienne soit environ 7 mois sur 12 (d'Avril à Octobre). Une poussière de Cantor ayant un tel générateur aurait une dimension égale à Log7/Log12 soit 0.783, valeur très proche de celle de la dimension que nous avons déterminée empiriquement.

Cette présentation ne doit cependant pas nous faire oublier qu'il y a dans notre définition de l'occurrence de pluie une notion de seuil qui renvoie bien sûr à celle d'intensité. Nous avons représenté sur la même Fig. 4 les ajustements obtenus par la méthode du comptage de boîtes en choisissant des seuils croissants (20, 40, 60 et 80 mm) pour définir l'occurrence de pluie. On constate que s'il est à chaque fois possible d'estimer une dimension fractale pour l'occurrence de pluie, celle ci dépend du seuil choisi, et que la dimension

décroit lorsque le seuil augmente (Hubert, Friggit et Carbonnel, 1990). Cette dépendance de la dimension d'un ensemble à la valeur de son seuil de référence, déjà notée par exemple par Schertzer et Lovejoy (1984), a amené de nombreux auteurs (Halsey et al, 1986; Schertzer et Lovejoy, 1987) à rejeter en pratique dans ce type d'étude la notion d'objet fractal au profit de celle de champ multifractal. Le phénomène étudié est alors représenté par une hiérarchie de fractals correspondant aux régions, de plus en plus ténues, sur lesquelles le champ dépasse des seuils de plus en plus élevés.

L'équation fondamentale des champs multifractals universels (Schertzer et Lovejoy, 1991; 1992), issue de la théorie des cascades multiplicatives, peut s'écrire, exprimée en termes probabilistes:

$$Prob[\varepsilon_\lambda > \lambda^\gamma] = K_\lambda \lambda^{-C(\gamma)} \qquad (4)$$

λ est le rapport de la plus grande échelle d'intérêt à l'échelle d'homogénéité considérée, ε_λ est l'intensité du champ à l'échelle λ, et γ est un ordre de singularité.

Cette équation asymptotique (pour $\lambda \to \infty$) traduit de quelle manière sont liées la distribution de probabilité d'apparition des singularités d'ordre supérieur à γ et la fraction de l'espace occupée par ces dernières exprimée par sa codimension $C(\gamma)$ (La codimension C d'un objet de dimension fractale D inclus dans un espace de dimension E est le complément à E de D, $C = E - D$). Le facteur K_λ apparaissant au second membre de l'éq. (4) est une fonction de λ de type logarithmique. Schertzer et Lovejoy (1991) ont montré que pour des processus multiplicatifs à flux conservatif la fonction de codimension $C(\gamma)$ ne dépend que de deux paramètres fondamentaux C_1 et α.

$$C_{(\gamma)} = C_1 \left(\frac{\gamma}{C_1 \alpha'} + \frac{1}{\alpha} \right)^{\alpha'} \quad avec \quad \frac{1}{\alpha} + \frac{1}{\alpha'} = 1 \qquad (5)$$

$C(\gamma)$ est la codimension des singularités d'ordre supérieur ou égal à γ. C_1 mesure l'écart à l'homogénéité (c'est à la fois l'ordre des singularités participant à la moyenne et la codimension de ces singularités $C(C_1) = C_1$) et α (index de Lévy compris entre 0 et 2) mesure l'écart à la monofractalité obtenue pour $\alpha = 0$.

Les paramètres C_1 et α peuvent être estimés à partir de données empiriques par diverses méthodes. Outre l'estimation des paramètres ces méthodes permettent d'apprécier les limites en temps du comportement scalant. Nous citerons la méthode PDMS ('Probability Distribution/Multiple Scaling') (Lavallée et al, 1991) et la méthode DTM ('Double Trace Moment') (Lovejoy et Schertzer, 1991), cette dernière méthode étant plus fiable car spécifique puisqu'elle permet une détermination du paramètre α indépendante de celle du paramètre C_1. Nous ne décrirons pas ici ces méthodes, mais nous signalerons que leur amélioration, la caractérisation de leur fiabilité et de la précision des estimations qu'elles fournissent, voire la définition de nouveaux algorithmes d'estimation restent des sujets majeurs de préoccupation pour les chercheurs.

Dans le domaine des précipitations de nombreuses estimations ont été réalisées utilisant diverses méthodes appliquées à des séries de localisation, de durée et de résolution très variées. Le comportement des séries apparait scalant pour des durées allant au moins de quelques minutes à l'année. On remarque que toutes les estimations du paramètre multifractal α sont voisines de 0.5 (Tessier et al, 1988; Hubert et al, 1993). Ce fait est de grande importance car dans ce cas, qui semble donc être la règle pour le champ temporel de précipitation, l'ordre des singularités du champ est borné et sa valeur maximale est:

$$\gamma_0 = -\frac{C_1\alpha'}{\alpha} = \frac{C_1}{1-\alpha} \qquad (6)$$

comme on peut l'établir à partir des éqs. (4) et (5).

L'expression de la codimension $C(\gamma)$ peut alors être reformulée en fonction des paramètres γ_0 et $C_0 = C(0) = C_1/\alpha^{\alpha'}$

$$C(\gamma) = C_0\left(1 - \frac{\gamma}{\gamma_0}\right)^{\alpha'} \qquad (7)$$

Les notions d'ordre de singularité peuvent paraître étrangères aux préoccupations habituelles des hydrologues. C'est pourquoi nous allons transformer nos équations pour revenir à des notions familières: les hauteurs d'eau accumulées sur une période et les probabilités correspondantes. Partant de l'équation fondamentale (4) où ε_λ est une intensité, nous définirons une accumulation à l'échelle λ comme $H_\lambda = \varepsilon_\lambda/\lambda$ et il vient alors:

$$Prob[H_\lambda \geq \lambda^{\gamma-1}] = K_\lambda \lambda^{-C(\gamma)} \qquad (8)$$

Pour une probabilité au dépassement donnée p, nous rechercherons la valeur $H_\lambda(p)$ telle que:

$$Prob[H_\lambda \geq H_\lambda(p)] = p \qquad (9)$$

Nous avons alors

$$p = K_\lambda \lambda^{-C(\gamma)} \quad soit \qquad C(\gamma) = (Log K_\lambda/p)Log(\lambda)$$

$$H_\lambda(p) = \lambda^{\gamma-1} \quad soit \quad Log H_\lambda(p) = (\gamma-1)Log\lambda$$

En inversant l'expression (6) il vient:

$$\gamma = \gamma_0\left[1 - \left(\frac{C(\gamma)}{C_0}\right)^{1/\alpha'}\right] \qquad (10)$$

et donc

$$Log[H_\lambda(p)] = -\gamma C_0^{-1/\alpha'}[Log\lambda]^{1/\alpha}\left[Log\left(\frac{K_\lambda}{p}\right)\right]^{1/\alpha'}$$

$$+ (\gamma_0 - 1)Log\lambda \qquad (11)$$

Cette relation est d'une grande importance théorique et pratique. Il s'agit en effet d'une relation générale, *invariante d'échelle*, entre fréquence (p), durée (repérée par le facteur d'échelle λ) et hauteur de précipitation ($H_\lambda(p)$).

Nous examinerons d'abord une conséquence fondamentale de cette relation. Si α' est négatif, ce qui est le cas lorsque $0 < \alpha < 1$ et nous avons noté plus haut que toutes les estimations disponibles de α étaient voisines de 0.5, le premier terme du second membre de l'expression (11) tend vers 0 avec p, et on obtient donc à la limite:

$$Log[H_\lambda(0)] = (\gamma_0 - 1)Log\lambda \qquad (12)$$

Dans ces conditions, $H_\lambda(0)$, hauteur de précipitation dont la probabilité au dépassement est nulle, est donc *finie*. Il existe pour H_λ une valeur maximale ne pouvant être dépassée, c'est à dire une précipitation maximale possible.

De nombreux manuels d'hydrologie (Chow, 1964; Réménieras, 1965; Raudkivi, 1979) proposent une figure rassemblant les records de pluie enregistrés de par le monde pour différentes durées allant d'une minute à deux ans (Fig. 5). Portés sur un diagramme Log-Log les points représentatifs de ces records s'alignent de façon satisfaisante selon une droite de pente voisine de 0.5. On peut penser que l'éq. (12) ouvre quelques perspectives pour expliquer ce remarquable alignement. Bien plus, les paramètres que nous avons estimés ($\alpha = 0.5$ et $C_1 = 0.2$) pour une série (d'une durée de 1 an et de résolution 6 minutes) recueillie sur l'île de la Réunion où ont été enregistrés les records pour des durées allant de quelques heures à quelques jours permettent d'estimer une pente théorique de la droite des records égale à 0.6 proche du 0.5 observé.

Une autre remarque peut être formulée à propos de l'éq. (11), qui ouvre la voie à sa validation et à son utilisation pratique. Pour une échelle de temps donnée (caractérisée par le rapport d'échelle λ),

$$Log[H_\lambda(p)] \quad est \ une \ fonction \ linéaire \ de \ \left[Log\left(\frac{K_\lambda}{p}\right)\right]^{1/\alpha'}$$

$$(13)$$

dont l'ordonnée à l'origine n'est autre que Log $[H_\lambda(0)]$, la précipitation maximale possible à l'échelle de temps considérée.

Etant donnée une série empirique de mesures de hauteurs de précipitations selon un pas de temps régulier quelconque, on peut tout d'abord estimer le paramètre α grâce par exemple à la méthode DTM. Nous pouvons ensuite attribuer à chaque hauteur de pluie une probabilité empirique au dépassement (cette opération peut être délicate, les formules classiques comme celle de Weibull étant peu adaptées à des distributions très asymétriques). Le paramètre K_λ peut lui aussi être estimé à partir des données empiriques. En effet, partant de l'éq. (8) et faisant tendre γ vers moins l'infini, en

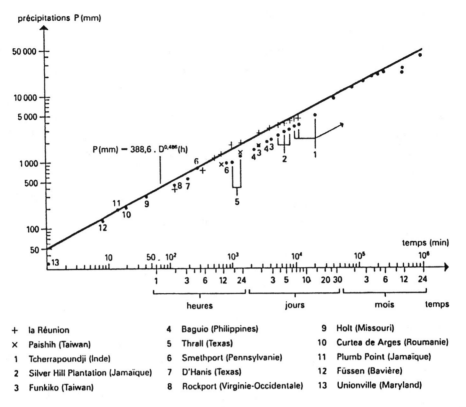

Fig. 5 Valeurs maximales de précipitations observées pour différents intervalles de temps. Les données s'ajustent à l'équation de régression $P = 50.54.D^{0.5}$, P étant exprimé em millimètres et D en minutes (Fig. reproduite de Réménieras et Hubert, 1990; d'après Raudkivi, 1979; données de Jennings, 1950; Paulhus, 1965).

remarquant que $C(\gamma)$ tend alors vers 0 puisque $0 < \alpha < 1$, K_λ peut être interprété comme la probabilité que la hauteur précipitée pendant l'intervalle de temps considéré soit *strictement* positive (compte tenu de la sensibilité des appareils de mesure, cette estimation peut se révéler très imprécise, surtout lorsque l'on considère de grands intervalles de temps).

Ayant estimé α et K_λ pour la série, et ayant attribué une probabilité au dépassement à chaque hauteur de pluie observée, il est possible de représenter ces couples probabilité-hauteur sur un diagramme par des points de coordonnées:

$$abcisse\ X = \left[Log\left(\frac{K_\lambda}{p}\right)\right]^{1/\alpha'}\quad ordonnée\quad Y = Log[H_\lambda(p)]$$

$$(14)$$

On peut alors apprécier l'alignement des points expérimentaux prévus par l'éq. (11) et estimer la précipitation maximale possible comme ordonnée à l'origine de sa droite représentative.

De premières applications concernant des séries dont le pas de temps va de 6 minutes à la journée ont été entreprises et sont encourageantes car elles permettent d'obtenir des alignements satisfaisants. Les effets des incertitudes entachant les estimations des paramètres α et K_λ d'une part, des

probabilités empiriques au dépassement d'autre part, doivent cependant faire l'objet de recherches complémentaires afin d'être bien compris et maîtrisés, particulièrement dans la perspective d'attacher un intervalle de confiance à l'estimation de la pluie maximale possible.

Il existe à l'heure actuelle deux points de vue opposés concernant les pluies extrèmes. Une école de pensée se réclame de la 'pluie maximale probable ou possible' (PMP); le principal mérite de cette approche est de s'appuyer sur une analyse physique des processus dont les aspects météorologiques, orographiques, etc.. sont pris en compte, mais on lui reproche souvent de demeurer trop qualitative particulièrement pour son application à l'aménagement. L'autre école de pensée s'appuie sur l'analyse statistique. Les hauteurs des précipitations sont considérées comme des variables aléatoires, leur succession comme un processus stochastique, sans aucune référence aux processus physiques sous jacents; on recherche alors essentiellement un bon ajustement des données empiriques à des lois plus ou moins arbitrairement choisies et en dépit de nombreuses incertitudes cette approche permet d'attribuer aux événements des probabilités fort appréciées des ingénieurs.

Le modèle de cascade multiplicative sur lequel nous nous appuyons est une schématisation simple des processus pluviogéniques, qui demande à être confortée et précisée. Cette schématisation se traduit, grâce à l'éq. (4), en termes statistiques et notre raisonnement permet finalement de définir explicitement la pluie maximale possible comme la précipitation dont la probabilité au dépassement est nulle.

Nos premiers résultats, outre leur intérêt propre qui est de proposer une relation générale *invariante d'échelle* entre hauteur, fréquence et durée, laissent entrevoir la possibilité d'une réconciliation et d'un dépassement des deux points actuellement développés à propos des pluies extrêmes.

BIBLIOGRAPHIE ————————

BARRETT E.C., MARTIN D.W. (1981) The Use of Satellite Data in Rainfall Monitoring, Academic Press, 340 p.

CHOW V.T., Editor (1964) Handbook of applied hydrology, Mc Graw Hill.

HALSEY T.C., JENSEN M.H., KADANOFF L.P., PROCACCIA I., SHRAIMAN B.I. (1986) Scaling measures and singularities, Phys. Rev. A, 33, p 1141.

HENTSCHEL H.G.E., PROCCACIA I. (1983) The infinite number of generalised dimensions of fractals and strange attractors, Physica, 8D, pp 435–444.

HUBERT P., CARBONNEL J.P. (1988) Caractérisation fractale de la variabilité et de l'anisotropie des précipitations intertropicales, CRAS, 307, série II, pp 909–914.

HUBERT P., CARBONNEL J.P. (1989) Dimensions fractales de l'occurrence de pluie en climat soudano-sahélien, Hydrol. continent., 4, pp 3–10.

HUBERT P., FRIGGIT F., CARBONNEL J.P. (1995) Multifractal structure of rainfall occurrence in West Afica, New Uncertainty Concepts in Hydrology and Water Resources (Z.W. Kundzewicz editor), Cambridge University Press, pp 109–113.

HUBERT P., TESSIER Y., LOVEJOY S., SCHERTZER D., SCHMITT F., CARBONNEL J.P., VIOLETTE S., DESUROSNE I. (1993) Multifractals and extreme rainfall events, Geophysical Research Letters, 20, pp 931–934.

HUDLOW M.D., PATERSON V.L. (1979) GATE Radar Rainfall Atlas, N.O.A.A. Special Report, U.S. Department of Commerce, Wasington D.C.

JENNINGS A.H. (1950) World's greatest observed point rainfall, Monthly Weather Rev., 78, pp 4–5.

LAVALLÉE D., SCHERTZER D., LOVEJOY, S. (1991) On the determination of the codimension function, Non-linear variability in geophysics, Kluwer, Dordrecht, pp 99–110.

LEBEL T., SAUVAGEOT H., HOEPFFNER M., DESBOIS M., GUILLOT B., HUBERT P. (1992) Sahelian Rainfall Estimation: The EPSAT-Niger Experiment, Hydrological Sciences-Bulletin- des sciences hydrologiques, 37, pp 201–216.

LOVEJOY S., SCHERTZER D. (1986) Scale invariance, symmetries, fractals and stochastic simulations of atmospheric phenomena, Bulletin of the AMS, 67, pp 21–32.

LOVEJOY S., SCHERTZER D., TSONIS A.A. (1987) Functional box-counting and multiple elliptical dimensions in rain, Science, 235, pp 1036–1038.

LOVEJOY S., SCHERTZER D. (1991) Multifractal analysis techniques and the rain and clouds fields from 10^{-3} to 10^6 m, Non-linear variability in geophysics, Kluwer, Dordrecht, pp 111–144.

MANDELBROT B.B. (1975) Les objets fractals, forme, hasard et dimension, Flammarion, Paris, 190 p.

MANDELBROT B.B. (1982) The fractal geometry of nature, Freeman, San Francisco, 461 p.

PAULHUS J.L.H. (1965) Indian Ocean and Taïwan rainfall set new records, Monthly Weather Rev., 93, pp 331–335.

RAUDKIVI A.J. (1979) Hydrology, Pergamon Press, Oxford New-York, 479 p.

RÉMÉNIERAS G. (1965) L'hydrologie de l'ingénieur, Eyrolles, Paris, 456 p.

RÉMÉNIERAS G., HUBERT P. (1990) Article 'Hydrologie' de l'Encyclopaedia Universalis, Paris, volume XI de l'édition 1990, pp 796–806.

SCHERTZER D., LOVEJOY S. (1984) On the dimension of atmospheric motions, Turbulence and chaotic phenomena in fluids, T. Tatsumi ed., North-Holland, pp 505–508.

SCHERTZER D., LOVEJOY S. (1987) Physical modeling and analysis of rain and clouds by anisotropic scaling and multiplicative processes, J. Geophys. Res., 92, D8, pp 9693–9714.

SCHERTZER D., LOVEJOY S. (1988) Multifractal simulations and analysis of clouds by multiplicative processes, Atmospheric Research, 21, pp 337–361.

SCHERTZER D., LOVEJOY S. (1989) Generalised scale invariance and multiplicative processes in the atmosphere, Pageoph, 130, pp 57–81.

SCHERTZER D., LOVEJOY S. (1991) Nonlinear geodynamical variability: multiple singularities, universality and observables, Non-linear variability in geophysics, Kluwer, Dordrecht, pp 41–82.

SCHERTZER D., LOVEJOY S. (1992) Scale, chaos and precipitation (ce volume).

TESSIER Y., LOVEJOY S., SCHERTZER D. (1988) Multifractal analysis of global rainfall from 1 day to 1 year, Nonlinear Variability in Geophysics 2, abstract volume, Paris.